ARITHMETICAL WORD PROBLEM SOLVING AFTER FRONTAL LOBE DAMAGE

ARITHMETICAL WORD PROBLEM SOLVING AFTER FRONTAL LOBE DAMAGE

A cognitive neuropsychological approach

Luciano Fasotti

Institute for Rehabilitation Research, Hoensbroek

SWETS & ZEITLINGER B.V. AMSTERDAM / LISSE

PUBLISHING SERVICE

The Institute for Rehabilitation Research (IRV) was established in 1981 and has close ties with the University of Limburg, Maastricht; The Lucas Foundation for Rehabilitation, Hoensbroek; and the Dutch Organisation for Applied Scientific Research TNO. The research conducted at the IRV is multidisciplinary in nature. It builds on the interface between the various medical and paramedical disciplines which are involved when addressing problems related to the individual's disability.

The IRV Series in Rehabilitation Research covers topics from the IRV's research programme, including studies on the rehabilitation process, augmentative communication and technology for independent living.

Library of Congress Cataloging-in-Publication Data

(applied for)

Cip-gegevens Koninklijke Bibliotheek, Den Haag

Fasotti, L.

Arithmetical word problem solving after frontal lobe damage : a cognitive neuropsychological approach / L. Fasotti. - Amsterdam [etc.] : Swets & Zeitlinger.– (IRV Series in rehabilitation research, ISSN 0925-8396 ; vol. 4)
Ook verschenen als proefschrift Maastricht, 1992 - Met lit. opg., reg. - Met samenvatting in het Nederlands.
ISBN 90-265-1308-9
NUGI 742
Trefw.: hersenletsel / neuropsychologie

Cover design: Rob Molthoff
Typeset: Athmer/SKP
Cover printed in the Netherlands by Casparie, IJsselstein
Printed in the Netherlands by Offsetdrukkerij Kanters B.V., Alblasserdam

ISSN 0925 8396
ISBN 90 265 1308 9
NUGI 742

Aan Alessandro

Contents

CHAPTER I

Introduction

1.1. Neuropsychology and Arithmetical Word Problem Solving

The interest in arithmetical word problem solving after frontal lobe damage is not new in neuropsychology. In 1967, Luria & Tsvetkova published a penetrating study on the subject. The conclusions of this study can be summarized in what follows. Patients with frontal lobe lesions can carry out additions, subtractions, multiplications and divisions without any particular difficulty. This suggests that they have no primary disturbance in the execution of arithmetical operations. Severe disturbances occur when arithmetic word problems consist of a series of successive, mutually dependent problem conditions. With these complex word problems, the frontal patient is no longer able to generate and execute a correct solution plan. The problem arises, so Luria & Tsvetkova argue, because the patient omits any preliminary investigation into the conditions of the problem and often performs an arithmetical operation on the first fragment of the problem that catches his attention.

These features differ markedly from the disturbance displayed in word problem solving by patients with left posterior (parieto-occipital or parieto-temporal) brain lesions. According to Luria & Tsvetkova, the latter patients repeatedly try to analyse the conditions of the problem. However, these attempts fail because the left posterior patients' processing of sentences containing complex logical grammatical expressions, such as prepositions and conjunctions, is disturbed. Such a "semantic" aphasia is not present in the frontal patient. The difficulties of frontal patients described by Luria & Tsvetkova were also observed by Lhermitte et al. (1972) and Christensen (1975). Unfortunately, all these studies were based on single-case observations and were mainly descriptive. At the end of the seventies and throughout the eighties, however, very little has been published on arithmetical word problem solving in the neuropsychological literature. In the area of cognitive psychology on the other hand, there has been a large surge of interest in the matter. Recent cognitive analyses of mathematical word problem solving, of which arithmetical word problem solving is part, have provided new insights in the mental processes involved in word problem solving (Riley et al., 1983; Mayer et al., 1984; De Corte & Verschaffel, 1987). Although a comprehensive theory of mathematical word problem solving is still lacking, some fundamental cognitive processes underlying word problem solving have been investigated and described. A general finding is that the first main stage in the solution of mathematics word problems consists of a complex text-processing activity. This text-processing is aimed at encoding the problem-text and can be broken down into two processes. Firstly, an analytical process, in which a literal phrase-by-phrase translation of each sentence of the problem into a memory representation takes place. Secondly, a largely synthetic process, in which the solver tries to integrate the information of each sentence into a familiar pattern or a "schema" of the problem. The second main stage refers to the search of the problem space in memory for a solution. In this stage the problem solver establishes a solution plan based on his knowledge of appropriate solving strategies, and performs arithmetical calculations using his knowledge of arithmetical algorithms.

The cognitive approach to the issue of arithmetical word problem solving offers some new perspectives for the investigation of defective word problem solving in brain damaged subjects. It may be useful in determining which problem solving processes are impaired after brain lesions with different localizations. Moreover, knowledge of impaired cognitive processes can facilitate the planning of remedial interventions.

1.2. Overview of this Study

Chapter II describes the effects of frontal lobe damage on human behavior and on arithmetical word problem solving in particular. The effects that frontal lobe damage can have on motor skills, on several cognitive functions, and on personality are briefly reviewed. In a separate section the most prominent theories on the specific function of the frontal lobe in cognition are examined. Subsequently, the differences between acalculia and arithmetical word problem solving impairments are outlined. A comprehensive description of what is known about impaired arithmetical word problem solving after frontal lobe damage concludes chapter II.

In chapter III several possible approaches to the issue of arithmetical word problem solving after frontal lobe damage are introduced and critically examined. It is explained why a psychometric approach cannot provide a satisfactory answer to the problem of defective arithmetical word problem solving in brain damaged subjects. A short presentation of the viewpoint of the "theory of activity" on problem solving issues is followed by a detailed description of the most comprehensive theory of arithmetical word problem solving after frontal lobe damage, namely that of Luria & Tsvetkova. In the evaluation of this theory, it is suggested that some basic concepts of the "theory of activity", used by Luria & Tsvetkova to explain impaired word problem solving, are ambiguous and poorly operationalized. Last but not least, the cognitive approach to problem solving is introduced. Some basic contributions of this approach to the issue of problem solving are explained. After that, an important shift of emphasis within the cognitive approach to problem solving is described. In the sixties and the seventies, cognitive analyses stressed the importance of general problem solving methods as a basis for skillful problem solving. A shift of emphasis has taken place in the eighties, when the role of domain-specific knowledge in problem solving became obvious. Within this last framework, much research was done on the nature of the knowledge required for mathematical word problem solving. This research has led to the stage analysis mentioned above. As the cognitive approach to mathematical word problem solving owes much to this stage analysis, a short outline of the evolution from general problem solving methods to domain-specific knowledge theories is presented. Ignorance of these developments has most probably led many neuropsychologists to neglect the subject of arithmetical word problem solving for such a long time. Even recent publications on mathematical disabilities (Deloche & Séron, 1987) do not take notice of arithmetical word problem solving. Another reason for this disregard might be the controversial nature of the study of problem solving in general. Even today, doubts are cast upon the possibility to study complex problem solving scientifically (Hussy, 1985). A brief historical introduction in chapter III illustrates the long-standing character of this issue in psychology. Finally, the cognitive approach to arithmetical word problem solving after brain damage is advocated. Two cognitive processes, namely the search for a solution strategy and the execution of arithmetical calculations, have already been comprehensively investigated in neuropsychological research. Therefore, the scope of the present study is limited to the stage of problem encoding, and in particular to the processes of sentence translation and integration of problem-information.

The subsequent chapters describe a series of experiments in which both these processes, i.e. sentence translation and information-integration, were compared in healthy controls, frontal and left posterior-injured patients. Chapter IV describes a study on the translation of several types of word problem sentences to an internal representation. This translation process was studied by means of a recognition task and a sentence-picture matching task. In addition, the relationship between sentence representation and arithmetical word problem solving was investigated.

In chapter V the recognition of problem-schemata or word problem patterns was examined in the same three groups. For this purpose, an arithmetical word problem sorting task was presented to each subject. The implications of different kinds of sorting behavior for arithmetical word problem solving are discussed.

In chapter VI an attempt was made to improve the encoding of word problem sentences and word problem-schemata in frontal and left-posterior brain damaged patients. A cognitive cueing procedure was applied in order to see if the problem solving performance of both groups could be improved by better encoding. Moreover, the consistency of this improvement was assessed over time.

In the final chapter, the research findings are discussed in the context of several other viewpoints on problem solving after frontal lobe damage. Several suggestions for further research in arithmetical word problem solving after frontal lobe damage are presented. Subsequently, a plea in favour of tasks with a high predictive value for everyday problem solving in neuropsychological research is made.

1.3. The Value of Arithmetic Word Problems

The choice of arithmetic word problems as a subject of study in brain damaged subjects was based on two considerations. Although in the present study some domain-specific skills in arithmetical word problem solving are emphasized, there is no doubt that several arithmetical word problem solving abilities are useful in many other problem solving tasks. In other words, arithmetic word problems are quite representative for general problem solving and intelligence. This is underlined by the fact that such problems are typically used in almost every test of general intelligence and special aptitude. According to Luria & Tsvetkova, it is quite obvious that the description of the solving process of arithmetic word problems contains the characteristic features of almost every intellectual activity, such as preliminary orientation, planning, execution of the plan and control. Moreover, arithmetic word problems have a potential value as representations of real-world situations (Resnick & Ford, 1981). In other words, arithmetical word problem solving may offer the opportunity to apply arithmetical knowledge in a wide range of real-life situations, such as shopping, travelling and administering finances. What neuropsychological research reveals about impaired arithmetical processes in brain-damaged subjects facilitates the design of training programmes aimed at remediating these impairments. On the long term such interventions should contribute to the functional independence of patients with brain lesions.

CHAPTER II

Frontal Lobe Damage

In this chapter, an attempt is made to describe the effects of frontal lobe damage on human behavior in general, and on arithmetical word problem solving in particular. First, several behavioral alterations produced by frontal lobe lesions are outlined. Then, the issue of the specific function of the frontal lobe in cognition is examined. Finally, the difficulties that patients with frontal lobe lesions experience in solving arithmetic word problems are described.

2.1. Disturbance of Cortical Functions after Frontal Lobe Damage

The consequences of frontal lobe damage for brain functioning have always been among the more complex problems in neurology as well as in neuropsychology. This is not only due to the complex anatomical structure of the frontal lobe (Luria, 1969), but also to some more technical issues. Laboratory techniques such as computerized tomography and electro-encephalography are not always sufficiently sensitive to the location and extent of brain pathology (Stuss & Benson, 1986), whereas psychometric tests often cannot assess specific forms of frontal lobe damage (Shallice, 1982; Wang, 1987). Furthermore, a multiplicity of functions is involved in frontal lobe regulation and control. As a consequence, damage to the frontal lobes, unlike damage to the three other lobes, does not result in a very specific clinical syndrome with clear-cut neuropsychological symptoms (Bigler, 1988).

Impairments after frontal lobe damage can be grossly subdivided into three categories: motor deficits, personality changes and cognitive impairments. The first two of these categories will be shortly described in the following sections, the latter more extensively in a separate paragraph.

2.1.1. Motor Deficits

Of all frontal lobe functions, the best known is the control over motor responses. The motor zones in the frontal lobe comprise three different regions: the primary motor area (Brodmann's area 4 in the precentral gyrus), the premotor area (Brodmann's areas 6 and part of area 8) and the prefrontal area (areas 45, 46 and 10). Figure 2-1 illustrates these areas.

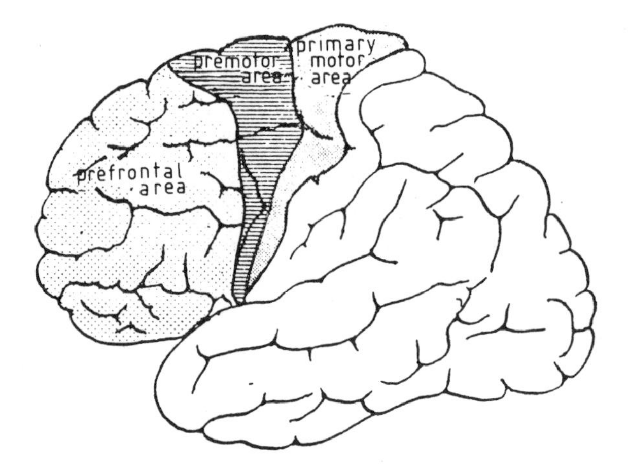

Fig. 2-1: Motor areas in the frontal lobe.

Motor, premotor and prefrontal zones perform different motor functions. The motor functions of the primary motor area have been extensively studied. In general, lesions of this area result in paralysis of paresis of the contralateral limb (for a review, see Henneman, 1980a, 1980b).

According to Luria (1969), the disturbance of movements arising from a lesion of the premotor area consists of two interrelated symptoms: deautomatization of complex motor acts and revival of elementary automatisms. Complex motor acts are composed of a number of independent impulses that must be coordinated into a single movement, the so-called "kinetic melody". Through long practice this synthesis becomes automatized; one impulse is sufficient to produce an entire movement. Injury to the premotor zone disturbs this automaticity; the speed and the smoothness of the complex motor act is lost and the individual components of the act require separate initiation again. A typist, for example, will loose the ability to type smoothly and will be forced to tap each letter separately. When the injury extends itself deeper and includes the basal ganglia, this can result in the compulsive repetition of an initiated action. Two examples of disturbed movements after premotor area damage are illustrated in Figure 2-2.

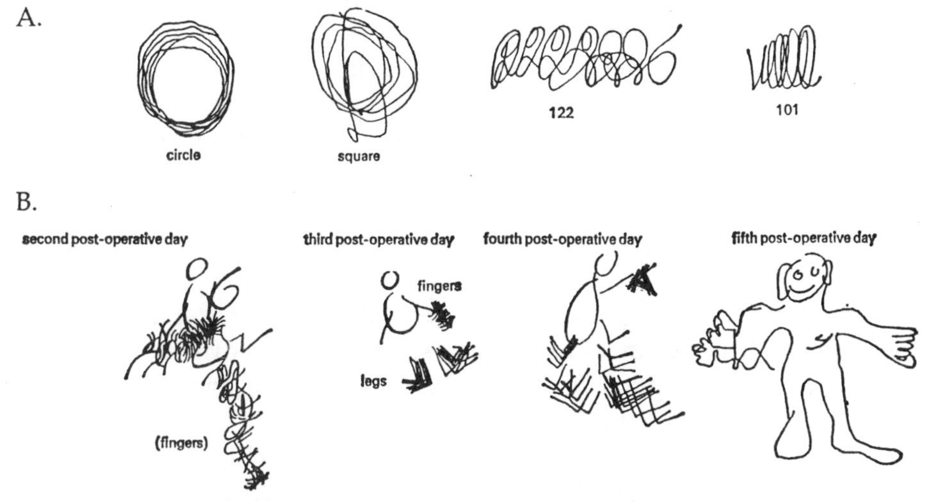

Fig. 2-2: **A. Disturbed movements in drawing and number-writing of a patient with a haematoma of the left premotor region (from Luria, 1973a).**
B. Drawing of a man after removal of a meningioma from the premotor area (from Luria, 1973a).

Lesions to the prefrontal area can also have direct effects on the regulation of movements. In this case, according to Luria (1966), the impairments include:

1. A difficulty to carry out direct motor instructions. An extreme distractability hinders the patient with prefrontal damage in the initiation of the required movement.
2. Although verbal commands are present in the patient's memory, their controlling function over motor behavior is lost. If told to tap once whereas the examiner taps twice and vice versa, the patient soon imitates the examiner, even if he still can correctly repeat the instructions of the task.
3. The replacement of complex motor sequences by simpler sequences or by inert stereotypes. The acts of striking a match and lighting a candle are

correctly performed, but later on, the candle will for example be held like a cigarette.

4. The lacking comparison of actual movements with the original intentions; consequently, frontal patients do not notice the defective execution of movements.

More recent single case and group studies of patients with frontal lesions (Drewe, 1975; Canavan et al., 1985; Leonard et al., 1988; Canavan et al., 1989) suggest that the syndrome is not as consistent as Luria describes it.

The combination of psychometric and post mortem evidence indicates that the prefrontal syndrome in question arises only when more global cerebral dysfunction through displacement, distortion, disrupted blood circulation or oedema is present (Canavan et al., 1985).

Moreover, some of the aforementioned symptoms have not only been observed after frontal but also after parietal (Kolb & Milner, 1981; De Renzi et al., 1983) and temporal damage (Leonard et al., 1988; Canavan et al., 1989).

2.1.2. Personality Changes

Alterations in emotional processes and personality are characteristic of frontal lobe damage. These changes have been reported for well over a century (see J.M. Harlow's description of P. Gage in Walsh (1978) and A. Brickner's description of patient A. in Damasio (1979)).

These case studies, together with other descriptions (Walsh, 1978; Damasio, 1979; Fuster, 1980), suggest that changes in personality after orbitofrontal and mediobasal lesions of the frontal lobe result in a coherent "frontal lobe syndrome". Patients suffering from this syndrome would be characterized by a lack of foresight, purposelessness, apathy and unconcern, impulsivity, sexual and personal hedonism, changes in attitude towards social and ethical rules, euphoria and facetiousness.

Other authors (Luria, 1969; Blumer & Benson, 1975; Lishman, 1978; Heilman & Valenstein, 1979; Stuss & Benson, 1984) suggest that frontal lobe lesions can produce two types of personality changes: a pseudodepressive syndrome caused by dorsolateral lesions, and a pseudopsychophatic syndrome, linked to basomedial lesions of the frontal lobe. The pseudodepressive syndrome is characterized by apathic-akinetic-abulic behavior. Unconcern, lack of drive, and lack of emotional reactivity are its principal symptoms. In the pseudo-psychopathic variant, disinhibitive phenomena like facetiousness, hedonism and lack of concern for others are predominant.

Other evidence for this distinction comes from Faust (1966). He reports on a patient in whom both syndromes occur at different stages in the development of a brain tumor. He concludes that

> "In this temporally split pathological progression the different symptomatologies of the dorsal and basal frontal lobe are reflected. Whereas the high frontal injuries produce loss of energy, affective dullness, loss of interest and complete mental incapacity for prospective plans, the basal injuries cause euphoric absence of the critic, loss of inhibitions, symptoms of moria and indifference to social habits" (p. 414-416).

A psychometric study by Maly & Quatember (1980) has confirmed the distinction between both syndromes. In this study two groups of patients, one with "frontoconvex" (dorsolateral) and another with "frontobasal" injuries, filled in a German personality inventory. The frontoconvex group scored high on Nervousness, Depression, Inhibition and Emotional Lability scales and had low scores on Sociability, Contentedness and Extraversion, whereas the frontobasal group was high on Nervousness, Agressiveness, Irritability, Contentedness, Extraversion and Emotional Lability, without low scores on any other scale.

Finally, in one of the few studies on the effects of lateralized frontal lesions on mood regulation, Grafman et al. (1986) conclude that patients with right orbitofrontal lesions tend to 'edginess'/anxiety and depression, whereas patients with left dorsolateral lesions seem to be more agressive and hostile.

2.1.3. *Cognitive Impairments*

2.1.3.1. *Sensory, Perception and Construction Functions*

The frontal lobes are widely recognized as motor areas, although they also subserve sensory-perceptive and constructive functions. This brief outline reviews the latter functions, as they appear after frontal lobe damage. Four major sections are covered: (1) hemi-inattention (2) visual search defects (3) visual-spatial disorders and (4) constructional disturbances.

Hemi-inattention or hemi-neglect can be defined as an inability to attend to stimuli in the sensory field contralateral to the side of a brain injury. There is strong experimental evidence, both from ablation and stimulation studies in animals, that the frontal lobes and/or their connections are involved in unilateral inattention phenomena (Bianchi, 1895; Welch & Stuteville, 1958; Crowne, 1983). Several studies have demonstrated that hemi-inattention can be caused by lesions in many interrelated areas of the brain, including the frontal cortex (Heilman & Valenstein, 1972; Watson et al., 1973; Damasio et al., 1980; Stein & Volpe, 1983).

Although inattention symptoms are usually caused by right hemisphere damage, neglect due to frontal lesions in the left hemisphere has repeatedly been reported (Curtis et al., 1972; Damasio et al., 1980).

Deficits in visual search are commonly observed after frontal lobe damage. These deficits have been explored by monitoring eye-movements during the patient's examination of thematic pictures (Luria et al., 1966). The meaning of these pictures eludes the patient, because he fails to scan the pictures systematically in search of meaningful details. Even when the experimenter asks the patient about relevant details, indirectly encouraging him to carry out the necessary scanning, the visual search process remains disorganized and haphazard.

Other authors, some of whom using other experimental tasks, have confirmed these findings (Teuber, 1964; Tyler, 1969; Guitton et al., 1982).

Visual-spatial disorders have been repeatedly evidenced after frontal lobe damage. The perception of reversible figures (e.g. the Necker cube) illustrates this deficit. Out of patients with different forms of brain damage, the group with unilateral frontal lobe damage perceives the fewest reversals (Teuber, 1964, 1966). The Aubert task, in which subjects are asked to align a luminous rod to the vertical in a dark room, produces a double dissociation between brain damaged patients. The group with frontal lobe damage makes the most exaggerated compensatory errors in the visual postural condition (with the body tilted and normal visual background), whereas parietal lobe patients are impaired in resetting the rod against a conflicting visual background (Teuber, 1964, 1966).

Visual-constructional disturbances after frontal lobe damage occur when an analysis of what is perceived, whether it are patterns or cube constructions, is necessary for reassembly. Simple block design tasks are performed correctly (McFie, 1969; Stuss et al., 1984), but in complex block design or construction, frontal patients are considerably impaired (Luria & Tsvetkova, 1964; Luria, 1966; Lhermitte et al., 1972; Walsh, 1978).

2.1.3.2. Attention

The brainstem-frontal system is often viewed as a unified and integrated complex for understanding attention. Disorders of attention produced by lesions of this system can subdivided into three major categories, depending on the neuroanatomical region involved; the reticular system, providing tonic levels of arousal and alertness, the diffuse thalamic projection system, responsible for phasic levels of alertness, and the fronto-thalamic gating system, related to selective and directed attention (Stuss & Benson, 1984).

The frontal area is closely connected to the reticular ascending system (RAS). Damage to this activating system results in arousal deficits, including difficulties to remain awake. In the worst case, the patient is in an akinetic-mute state; his sleep-wake cycles are normal, but there is little or no sign of cognitive or motor functioning. If phasic attention is intact, damage to the brainstem activating system may result in "drifting attention". This means that the patient can only be attentive for brief periods of time, after which he returns to his somnolent state (Benson & Geschwind, 1975).

"Wandering attention" is the opposite of drifting attention. The patient is alert, but he is easily distracted. This suggests a phasic disturbance of attention. This kind of attention disorder would result from damage to the the thalamic projection system (Benson & Geschwind, 1975).

Deficits in selective, directed attention may be caused by the frontothalamic gating system. Disorders of this type of attention frequently occur after focal frontal lobe damage. These manifest themselves in a marked inflexibility, a tendency towards perseveration (Milner, 1964; Nelson, 1976) and an increased distractability (Rylander, 1939; Luria, 1966). Shallice (1982) argues that this pattern of deficits can be described in terms of an impairment in attentional control. His interpretation of this phenomenon will be explained in what follows.

The relation between frontal lesions and impaired processes of sustained attention, already described in some earlier literature (e.g. Luria, 1973b), has also been emphasized in more recent studies (Salmaso & Denes, 1982; Wilkins et al., 1987).

2.1.3.3. Speech and Language

Frontal involvement in language functions has been generally assumed since Broca's demonstrations (1861, 1865). Damage to the frontal lobes can provoke several speech and language dysfunctions.

The most widely accepted of these language deficits is Broca's aphasia, also known as motor or non-fluent aphasia. This type of aphasia is associated with lesions of the left frontal lobe, which involve Brodmann's areas 44 and 45 (see Figure 2-3). It is characterized by non-fluent speech (telegraphic, agrammatic, dysprosodic and dysarthric), with relatively intact comprehension, impaired repetition and difficulties in naming, reading and writing.

Mohr et al. (1978) have demonstrated that the severity of Broca aphasia is related to the location and the extent of the lesion. The entire series of symptoms occurs only when large lesions in the sylvian region, encompassing much of the operculum, the insula and the subjacent white matter are present. Less extensive and more superficial damage, on the other hand, results in only a few of the aforementioned symptoms (Damasio, 1981). A second form of aphasia in which the frontal lobes are crucial is transcortical motor aphasia (Stuss & Benson, 1984). This type of aphasia occurs after lesions of the areas anterior or superior to Broca's area (Figure 2-3).

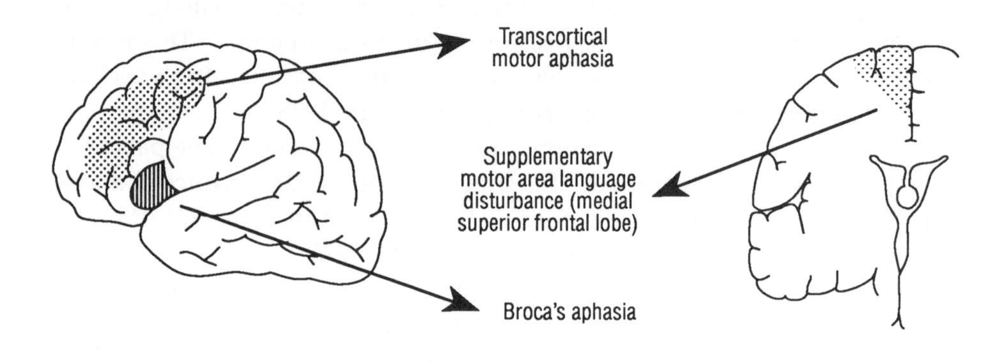

Fig. 2-3: Localization of different forms of language disturbance after frontal lobe damage (modified from Stuss & Benson, 1986).

The syndrome resembles Broca's aphasia; speech is non-fluent and agrammatical and comprehension is relatively unaffected. The key difference between transcortical motor aphasia and Broca'aphasia lies in the repetition of words and sentences. This repetition is intact in patients with transcortical motor aphasia (Goodglass & Kaplan, 1972), whereas it is impaired in patients with Broca's aphasia.

Damage to the frontal supplementary motor area (Figure 2-3) results in a syndrome characterized by the following symptoms: a decrease of verbal output and sometimes of facial expression and gesture, largely spared comprehension and repetition, and absence of paraphasia (Damasio & Van Hoesen, 1980; Ardila & Lopez, 1984).

Finally, recent research suggests that damage to the prefrontal parts of the brain, especially of the left hemisphere, can also impair language functions (Novoa & Ardila, 1987). Left dorsolateral lesions seem to disrupt the organization of linguistic information, i.e. the formation of the sequential pattern of utterances, whereas damage to orbitofrontal areas of the left prefrontal area produces confabulations and accidental associations disrupting the organization of the narrative (Kaczmarek, 1984).

2.1.3.4. Awareness

Several disorders of awareness, involving either the body or the environment, have been linked to frontal malfunction. One of these disturbances of awareness, hemi-inattention, has already been mentioned under the heading "Sensory, perception and construction functions".

The concept of anosognosia includes a large number of unawareness phenomena. Anosognosia can be defined as any state in which a patient denies an obvious physical abnormality. This denial has been first described by Babinski (1914) in several patients suffering from left hemiplegia. Although parietal lobe damage has been put forward as being the primary source, there are indications that frontal malfunction can play an important role in several unawareness phenomena (Stuss & Benson, 1986).

A connection between frontal lobe damage and confabulations has also been documented (Stuss et al., 1978).

Two other disorders entailing confabulations, i.e. reduplicative paramnesia and the Capgras syndrome, are also related to frontal lobe pathology. Reduplicative paramnesia refers to a reduplication of place; the patient knows the name of the hospital he is in, but locates it otherwise (Benson et al., 1976). In the Capgras syndrome, the reduplication involves persons. Individuals, usually members of the family, are believed to be impostors (Alexander et al., 1979).

Several other group studies have confirmed the existence of a correlation between unilateral or bilateral prefrontal lobe damage and confabulations or reduplicatory paramnesias (Mercer et al., 1977; Ruff & Volpe, 1981; Shapiro et al., 1981).

2.1.3.5. Memory

The role of frontal lobe functioning in memory has already been discussed in early animal experiments (Jacobsen, 1935, 1936). In these experiments, the performance of frontally lesioned monkeys was found to be poor in delayed response and delayed alternation tasks.

In delayed response tasks the experimenter places food under one of two containers and the animal learns to discriminate between the two containers. Having done so, the experimenter delays the animal's response by putting the containers out of reach or out of sight for some time. In delayed alternation, the trials are temporally interlocked with one another by making the delayed response depend upon the preceding response. Jacobsen concluded, with some hesitation, that the deficit underlying poor performance in these tasks was one of "immediate memory".

Since then, this hypothesis has been challenged by a number of authors, who have put forward a number of alternative explanations for impaired performance in delayed response and delayed alternation tasks. Increased distractability (Finan, 1942; Malmo, 1942; Bartus & Levere, 1977), hyperactivity and hyperreactivity (Orbach & Fischer, 1959; Buffery, 1967), deficits in habituation (Grueninger & Pribram, 1969) and task specific disturbances (Pribram, 1961) are some of the concepts used to explain defective performance in both the aforementioned tasks.

Several authors have demonstrated that frontal lobe pathology also impairs human memory functioning (Hécaen, 1964; Corkin, 1965; Ladavas et al., 1979; Milner, 1965, 1982). As in animal experiments, however, memory impairments have rarely been considered as a primary phenomenon but have often been attributed to secondary factors. Disrupted processes of selective attention and a lack of initiative seem to be the primary sources of apparent memory impairment after frontal lobe damage.

The disturbance in selective attention is apparent in the inability to deal with interfering information (Stuss et al., 1982). The lack of initiative is evident in the "forgetting to remember" phenomenon, that hinders the recall of elements that have not been entirely forgotten (Hécaen & Albert, 1978).

From the analysis of the memory disorder shown by a patient with bilateral frontal damage, Baddeley & Wilson (1988) conclude that the pattern of memory deficits after frontal lobe damage is not so much an entirely different form of amnesia, but rather a combination of a classic amnesic syndrome (e.g. impaired verbal and nonverbal episodic memory) and problems associated with a frontal dysexecutive syndrome: impaired procedural learning, inaccurate and slow performances in tests of semantic memory and confabulations in autobiographical memory. The latter symptoms do not occur in classic amnesics and are indicative of an impairment of the central executive component of working memory (Baddeley, 1986).

2.2. Specific Functions of the Frontal Lobe in Cognitive Functioning

Apart from the just discussed impairments in cognitive functioning, which may also occur after lesions to other than frontal parts of the brain, several theories have been proposed in order to explain cognitive dysfunctioning that may be dependent on the frontal lobe alone.

2.2.1. Explanations of Impaired Cognition after Frontal Lobe Damage

The problem of the specific effects of frontal lobe damage on cognitive functioning, although still not definitely resolved, has been partially clarified over time. Theories put forward to explain the specific effects of frontal damage on cognition usually also implicitely account for the fundamental incapacity of frontal patients to solve problems. Two prominent earlier theories are those of Goldstein (1936) and Halstead (1947).

According to Goldstein (1936, 1939), the cognitive functioning of patients with frontal lobe damage would be hampered by a basic lack of abstract behavior and by a pervading tendency toward concrete thinking. This loss of abstract thinking is, in a wide sense, defined as "lack of initiative, foresight, activity, and ability to handle new tasks." (Goldstein, 1944, p. 192).

Halstead (1947) uses a psychometric approach in an attempt to define biological intelligence, a type of intelligence that would represent "usable intelligence" in everyday life situations, rather than the more specific knowledge measured by standard intelligence tests. Halstead factor-analysed the results of 27 tests of brain damaged patients with various lesion locations. He concluded that, although the factors he had discovered were related to the whole cortex, they were mainly represented in the frontal lobe.

In more recent theories about frontal lobe damage, impairments in problem solving are referred to as disorders in executive functioning. Executive functions comprise the mental capacities involved in formulating goals, planning how to achieve these goals, and carrying out the plans effectively (Lezak, 1982).

Pribram (Miller et al., 1960) for example, has put forward that, instead of a reflex arch, the basic element of behavior would be a so called Test-Operate-Test-Exit (TOTE)-unit. In this TOTE-sequence, the feedback loop would play a fundamental role.

All behavior would follow from an organism's awareness of an incongruity between an actual stimulus-state and a desired state (TEST). This should lead to some kind of action (OPERATE) to reduce the incongruity. Again, the magnitude of the incongruity would be assessed (TEST). The elimination of the incongruity would end the whole sequence (EXIT) (see Figure 2-4). Information from the different tests would not act as a reinforcer, but would be used for comparative purposes.

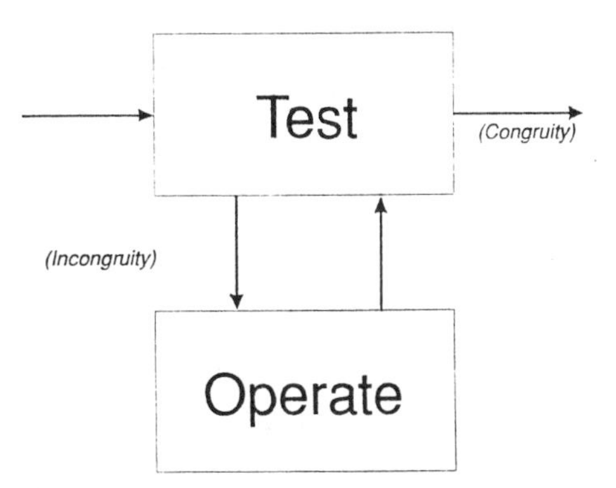

Fig. 2-4: The TOTE-unit (adapted from Miller et al., 1960).

A TOTE-unit can also be thought of as a larger operational unit, composed of different TOTE-subunits, each with its own feedback loop. The TOTE-system has two main functions: a planning function, which is carried out by a hierarchical structure controlling the construction as well as the sequencing of operations to be carried out; and an operating function, which consists in carrying out the plans. Planning is the internal representation of a task or a solution, whereas action by the organism is an external representation of the neural program in the brain (Pribram, 1967).

The TOTE-system can be outlined as a fourfold division of brain function (Miller et al., 1960). The processing and discrimination of information would be subserved by the sensory projection areas and the posterior association areas (the "external" parts of the brain), whereas planning functions would be operated by the limbic system and the frontal association areas (the "internal" portion of the brain). Behavioral plans could thus be formulated posteriorly and transferred to the frontal regions, where they await implementation.

Later, Pribram adds the factor of context to his theory (Pribram, 1973). He claims that the sequencing of behavior depends upon the context of events. After frontal lobe damage, behavior would lack context, because stored routines are inappropriate or missing. Pribram concludes that

> "the frontal cortex is especially concerned in structuring context-dependent behaviors." (p. 308).

Damasio (1979) stresses that the frontal lobe is concerned with high-level cognition. This author compares the functioning of the frontal lobes with and relates it to that of the hypothalamus. Through automatic responses to unbalancing factors, the hypothalamus controls the internal milieu so as to contribute to rewarding equilibrium (homeostasis). This homeostasis is also a function of environmental contingencies. In the case of man, these contingencies are regulated by complex rules of personal and objectal relationship.

> "In terms of this environment the mere evaluation of S and managing of R are not satisfactory and the need to override such quasi-automatic nonacquired responses has probably been answered by the development of complex decision chains...able to judge at progressively more elaborate levels, according to external as well as internal rules, and according to medium, long-term, and immediate goals." (p. 371).

To Damasio, the structures capable of performing this task reside in the frontal lobes. The frontal lobes would perform the same type of regulatory activity as the hypothalamus, but with hypercomplex environmental contingencies in the framework of the individual's own history, as well as in the perspective of his intended future course.

Another important principle in Damasio's theory is that he conceives the frontal lobe as

> "an outpost of the hypothalamus informed by learning rather than by inheritance." (p. 371).

He also hypothesizes that more and more courses of action are "taught" to non-frontal structures, so that frontal structures are cleared for further learning and programming.

Fuster (1980, 1985) emphasizes the critical role of the prefrontal cortex in the temporal organization of goal-directed behavior. Each sequence of behavior is a temporally structured series of actions, that requires the mediation of cross-temporal contingencies, not only between successive acts, but also between individual acts and the central representations of both the plan and the goal of behavior (see Figure 2-5). The latter contingencies are the most difficult to an organism engaged in novel or complex behavior. It is precisely for these contingencies that the frontal lobes would have a "time-bridging" function.

Evidence from research on lesion effects, neurochemical phenomena and metabolic activity leads Fuster (1985) to the conclusion that the prefrontal cortex subserves at least three cognitive functions: short-term memory, in order to permanently adjust actions to the original plan; anticipatory set, in order to adjust the same actions to the goal; and control of interference, to eliminate external and internal distractive sources that may disrupt the formation and execution of behavioral structures.

Shallice (1982) outlines an information-processing model which predicts that, after frontal lobe damage, the performance of non-routine tasks can be impaired, independently from the performance of routine tasks. His theory is built on three basic concepts: "schema", contention scheduling and the Supervisory Attentional System.

An action or thought schema is a basic unit that controls a specific overlearned action, like drinking from a container or finding one's way home from work. Schemas are often hierarchical and composed of a set of subordinate or lower-order schemas.

Contention scheduling is the process of activating the appropriate schemas necessary for routine behavior. These highly compatible schemas are activated by "triggers" from perception or from the output of other schemas.

In non-routine goal achievement (e.g. problem solving) contention scheduling is no longer sufficient and activation of appropriate additional schemas is needed. This additional activation is achieved by another mechanism -the Supervisory Attentional System (S.A.S.)- which contains the general programming or planning systems that can operate on schemas in every domain.

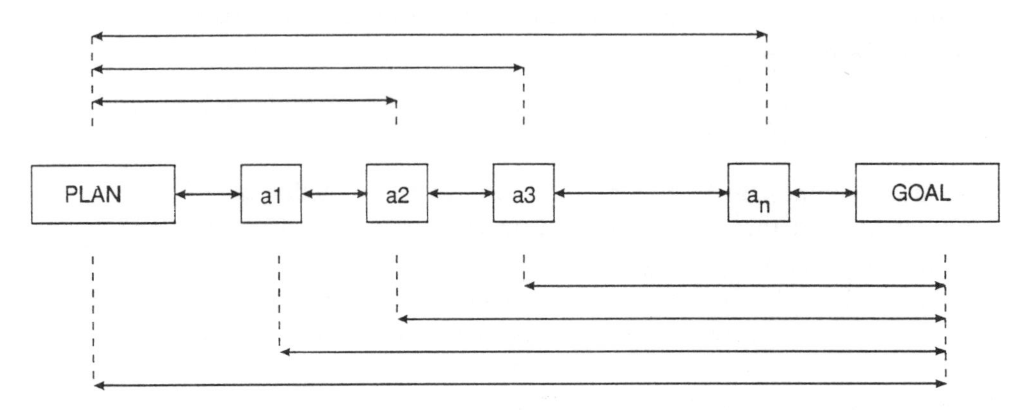

Fig. 2-5: Succession of acts (a_1...a_n) constituting a behavioral structure. Cross-temporal contingencies are represented by two-way arrows (from Fuster, 1985).

The Shallice model allows to make some predictions on high-level cognitive impairment, and in particular on what would happen after an impairment of the S.A.S.. Shallice postulates that S.A.S. is a function of the frontal lobe.

As the contention scheduling mechanism is not affected after frontal lobe damage, the performance on routine tasks should remain intact. Difficulties in dealing with novel tasks or with planned initiative would, however, occur. The model allows a couple of inferences to be made: one from contention scheduling properties, the other from S.A.S.-properties.

When S.A.S. is inoperative after frontal lobe damage, it leaves the organism operating under the control of contention scheduling only. If the environmental situation is such that there is a trigger present that strongly activates a schema, it will not be possible to prevent the schema from being selected. As a result, response perseveration should occur. If there are no triggers present in the environment, any new input can capture the process of contention scheduling and the brain-damaged patient will respond randomly and be easily distractable. According to Shallice, this combination of response perseveration and distractibility is clearly present in both animals and humans with frontal lobe damage.

The different theories on executive functioning have many features in common:

- they all stress the fundamental role of the frontal lobe in high-level, complex cognition concerned with *sequences* of actions
- they agree that the frontal lobe is concerned with novel, non-routine behavior
- they are at one in claiming that frontal lobe damage disrupts the planning

and programming capacities involved in complex, goal-directed behavior such as problem solving.

2.3. Frontal Lobe Damage and Arithmetical Word Problem Solving

2.3.1. *Arithmetical Word Problem Solving and Acalculia*

Before examining the issue of defective arithmetical word problem solving after frontal lobe damage, a brief discussion of the phenomenon of acalculia is necessary. Acalculia refers to the disruption of calculation skills after lesions in particular cerebral areas. Based on two large-scale studies (Hécaen et al., 1961; Hécaen & Angelergues, 1961), Hécaen has suggested a widely adopted classification of calculation defects. These defects are classified according to the presence of

1. difficulties in the spatial arrangement of numbers,
2. difficulties in the reading and writing of numbers, and
3. difficulties in the execution of arithmetical calculations.

A predominance of right and left posterior hemisphere lesions is found with the first type of calculation disorder, whereas left posterior lesions are found with the second and third types of acalculia.

Although calculation ability is an important element of arithmetical word problem solving, calculation deficits will not be addressed in the present study. The reason is that calculation disorders are not associated with lesions to the frontal lobes (Benton, 1987; Spiers, 1987).

2.3.2. *Defective Arithmetical Word Problem Solving after Frontal Lobe Damage*

As stated above, frontal lobe lesions do not appear to impair the execution of basic arithmetical operations such as addition, subtraction, multiplication or division. This absence of an outright acalculia after frontal damage has repeatedly been described (Luria, 1966; Christensen, 1975; Walsh, 1978). Even the phrasing of an arithmetical operation into a simple algorithm does not seem to be exceedingly difficult for the frontal lobe patient. For example, an arithmetic word problem of the type

Jack has 4 apples. Jill has 3 apples. How many apples have they in total ?

is readily solved by frontal patients.

With more complex word problems, however, defective problem solving performance may become apparent. Adding an intermediate step to the preceding problem, as for example in the following word problem:

> Jack has 4 apples. Jill has 2 more than Jack. How many apples have they in total ?

may elicit severe solving difficulties in the frontal patient (Stuss & Benson, 1986). In other words, arithmetic word problems consisting of several mutually dependent arithmetical operations and extending beyond the limits of a single arithmetical calculation are impaired in frontal patients.

Luria (1973a) has outlined two other types of arithmetic word problems in which frontal patients show severe disturbances. The first type of word problem presupposes the recoding of some of the givens, e.g.

> There are 18 books on 2 shelves; there are twice as many books on one shelf as on the other. How many books are there on each shelf ?

To solve this problem, apart from taking account of the given of "2 shelves", the solutor has to introduce the given of "3 parts" and calculate the number of books on each of these auxiliary "parts". What follows is an example of how this problem is solved by several frontal patients. They start by carrying out the operation 18:2=9, corresponding to the first sentence of the problem. This may be followed by the operation 18x2=36, corresponding to the proposition "there are twice as many books on one shelf", after which these patients believe that the problem has been solved.

The second type of problem that cannot be solved by frontal patients involves a conflicting condition that requires the inhibition of an impulsive response. An example of such a problem is:

> A pedestrian takes 30 minutes to reach the station, while a cyclist goes 3 times as fast. How long does the cyclist take ?

The impulsive response, common in frontal patients, is to answer "30x3=90 minutes", or something like it.

The characteristic disturbances which frontal patients experience with these complex arithmetic word problems are better described as difficulties in "verbal reasoning" (Luria, 1966) or "discursive intellectual activity" (Christensen, 1975). Instead of devising a step-by-step solution program, the frontal patient usually grasps only a particular fragment of the problem. With this fragment he starts to carry out isolated arithmetical operations. According to Christensen (1975) this way of proceding may transform the whole process of solution into a series of impulsive, fragmentary arithmetical operations, usually unconnected with the final goal of the problem. Only with external assistance that breaks the word problem into appropriate stages, asking for a response at each step, the frontal patient may achieve a correct answer.

In some cases this difficulty in taking into account all the givens of an arithmetic word problem is less evident, but the ensuing solution plan is incomplete. Thus, a frontal patient asked to solve the following word problem:

> A son is 15 years old. His father is 25 years older. His mother is 5 years younger than his father. How old are the 3 of them together ?

answers "The son is 15 years old, his father is 25 years older, that makes 40, and the mother is 5 years younger, that is 35.", without going any further, even when stimulated by the examiner (Lhermitte et al., 1972).

This example also characterizes a last impairment found in frontal subjects solving arithmetic word problems: the absence of a comparison between the obtained results and the initial givens of the problem. Even when frontal patients arrive at what appear to be ludicrous or odd answers, as in the just mentioned solution, manifest divergences between the givens and the final result are not corrected.

CHAPTER III

Cognitive Psychology and Arithmetical Word Problem Solving

3.1. Different Approaches to Problem Solving: Historical Perspective

The first approach to problem solving in experimental psychology was not an encouraging one. The founder of the discipline, Wilhelm Wundt (1832-1920), considered thinking, of which problem solving is part, to be a higher mental process. Wundt (1896) excluded such higher mental processes from proper experimental investigation and confined them to the less rigorous inquiries of the Völkerpsychologie, a descriptive study of collective thought processes.

Wundt had straightforward associationist views on the explanation of psychological phenomena and claimed that

> "... there is only one kind of causal explanation in psychology and that is the derivation of more complex psychical processes from simpler ones." (cited in Mandler & Mandler, 1964, p. 132).

Thinking, being a complex mental activity not preceded by any analysis of simpler mental processes, could not provide reliable and objective results.

Despite this negative attitude, Wundt's successors at Würzburg, led by Karl Marbe and Oswald Külpe, investigated thinking and problem solving. Their findings can roughly be grouped under two headings (Humphrey, 1951).

The first concerns the mechanism of thinking. According to the Würzburgers, thinking was not based on a passive association of ideas. It was a process led by "determining tendencies"; a mental set brought about by the formulation and the comprehension of task-instructions, that guided the later stages of solution.

The second important discovery was linked to the matter of thinking. The Würzburgers reported of "imageless thoughts" in thinking: conscious experiences that contained no trace of sensory or imaginal elements and that were inaccessible to further analysis by the subjects who experienced them.

The importance of the Würzburger Schule is closely related to these findings. The "determining tendencies" ran counter the empiricist view that thinking should only be an association of ideas, and "imageless thoughts" contradicted the ancient and widely accepted Aristotelian dogma that thinking is impossible without images.

Nevertheless, the controversy between the Würzburgers and their counterparts was never clearly settled. The reason is quite obvious:

> "It could hardly have been settled by experiment." (Johnson-Laird & Wason, 1977, p. 3).

In the period between the World Wars, problem solving remained a subject of controversy between associationist and non-associationist theories. The rise of behaviorism in the United States gave a new impulse to associationist explanations of problem solving issues.

The early behaviorists affirmed that problems are solved by trial and error, without much evidence of thinking. Thorndike (1898) for example, assumed that for a particular problem situation S there are many possible responses R1, R2, R3.... In addition, the actual link between problem situation and possible responses may vary, with some connections between S and R being very strong and others being very weak. Problem solving is simply trying out the most dominant response R1 and if it fails, trying out R2, and so on, until one works. This principle, Thorndike's "law of effect", states that responses which are not useful for the solution of a problem will loose strength and gradually disappear, whereas responses that lead to a correct solution are likely to be emitted more frequently.

Even in this form, associationist principles were consistently challenged by a group of German researchers: the Gestalt psychologists.

The most influential Gestalt opponent of Thorndike was undoubtedly W. Köhler. From his studies on the problem solving behavior of chimpanzees (1921), he concluded that problems are solved through a perceptual restructuration of the component elements of the problem. In one of his experiments a banana is placed somewhere on the outside of the cage of a chimpanzee. The banana is attached to a thread that ends inside the cage. It does not take the animal long before understanding the nature of the problem and pulling the banana into the cage. Köhler's interpretation was that suddenly the animal perceives the different elements of the problem as a meaningful, integrated whole.

Köhler's experiments illustrate two central concepts of the Gestalt theory of problem solving: structure and insight. The perception of integrated wholes (Gestalten) is a necessary condition for successful problem solving. Insight refers to the ability of the problem solver to perceive these structures, to grasp their mutual relations and to reorganize these relations in order to fill in missing elements.

Following Selz (1922), Gestalt psychologists also claimed that thinking can be productive. Applying old habits or behaviors to new problems, as in Thorndike's descriptions, is a matter of reproductive thinking. Thinking is productive when the problem solver creates new solutions to a problem or finds a new way to organize the elements that constitute the problem (Wertheimer, 1945).

Retrospectively, the different behaviors displayed by animals in Gestalt and behaviorist's experiments can, of course, be accounted for by the different task-demands involved in the problems set by the workers of the two different traditions.

The trial and error explanations of the early behaviorists may seem oversimplified and are virtually forgotten in the problem solving field today. On the other hand, behaviorists proposed a clear and testable theory, whereas Gestalt psychologists often tended towards vague and obscure formulations in order to define their concepts of structure and insight.

The fundamental differences between associationist and Gestalt views on problem solving issues are illustrated in Table 3-I.

Table 3-I: Fundamental differences between associationist and Gestalt views on problem solving.

	Associationist	*Gestalt*
1. Type of task	Reproductive	Productive
2. Mental activity	Try S-R links	Reorganize elements
3. Unit of thought	S-R links	Organizations
4. Detail of theory	Precise	Vague
		(based on Mayer, 1983)

This short historical survey would not be complete without mentioning the neo-associationist problem solving theories developed after World War II. Central to neo-associationist theories is the concept of mediation. According to Mayer (1983), mediation theories suggest that the S evokes a miniature internal response called mediational response or rm; the rm creates a new internal state or sm, this new sm may evoke a different rm, followed by a new sm, and so on, until finally an sm evokes an overt solution response R.

Two well-known mediational theories are those of Berlyne (1965), on thinking as a chain of symbolic internal responses, and of Osgood (1963), on mediating reactions as intermediates for the representation of language meanings.

3.2. Mathematical Ability and Impairments in Mathematical Problem Solving

The issue of mathematical ability and of defective mathematical problem solving, including impaired arithmetical word problem solving, is put by Mayer et al. (1984) in the following words:

> Why is it that some people, when faced with a mathematics problem, are able to generate clever solutions, whereas other people cannot ? What makes people differ in their performance on mathematics problems ? What is it that good mathematics problem solvers possess that poor mathematics problem solvers do not possess ? (p. 231).

Three basic approaches can be distinguished in answering this question.

3.2.1. The Psychometric Approach

The psychometric approach offers a straightforward solution to the issue of mathematical ability: subjects perform differently in mathematical problem solving because they have different amounts of mathematical ability at their disposal. Mathematical ability refers to the skill to solve mathematics problems efficiently (Mayer et al., 1984).

Investigators using psychometric methods have analysed patterns of individual differences in scores on various kinds of mental tests, among which tests of mathematical ability. The presupposition is that specific abilities give rise to the observed individual differences in test scores. The statistical procedure of "factor analysis", in which patterns of individual differences are analyzed in terms of hypothesized constructs or "factors", has often been used in the search for specific abilities (Spearman, 1927; Thurstone, 1938;

Guilford, 1967). Although the author is not aware of studies investigating impaired arithmetical word problem solving after brain damage with factor-analytical methods, such a procedure can readily be used for this purpose. It would be sufficient to factor-analyse the results of a variety of neuropsychological tests, in search of the specific "factors" underlying the poor word problem solving performance of several groups of brain damaged subjects with different sites of lesion.

In spite of this, the psychometric approach cannot provide a satisfactory answer to the problem of mental abilities in general and to the problem of mathematical ability in particular. Three main reasons for this setback are listed by Sternberg (1983):

1. The difficulty of psychometric methods in distinguishing among, and in empirically disconfirming alternative factorial theories, especially through the use of factorial methods.
2. The almost exclusive reliance of factorial methods upon individual differences for the identification of abilities, with the concomitant assumption that abilities only exist if they generate individual differences in task performance. In that line of reasoning, the psychometric definition of mental abilities in general and of mathematical abilities in particular must be considered as circular (Mayer et al., 1984).
3. The failure of factorial methods to identify the processes that constitute task performance. For the present study, aimed at a process-analysis of arithmetical word problem solving, a further perspective is needed. This perspective should shed more light on the underlying processes that constitute arithmetical word problem solving.

3.2.2. *The Theory of Activity*

3.2.2.1. *General Outline*

The "theory of activity" is a general theory of psychology developed in Russian psychology. In this theory human activity, that is the way that humans actively interact with one another and with the physical world around them, is seen as the basic category from which all psychological phenomena derive. This assumption implies, among other things, that

> "...mental activity is the result of transferring external, material actions to the plane of perception, representation, and concepts." (Galperin, 1959, cited in Wertsch, 1981, p. 251).

This transfer is realized through a multi-stage mechanism called "internalization". Internalization has been extensively studied by P.J. Galperin (1969), especially within the field of problem solving. Galperin distinguishes between two basic functions of a mental action in problem solving; an orienting function and an executive function. The orienting function of a mental action is its most important one. According to Galperin (1968)

> "It defines the outline of each operation and guarantees control of the action in the process of execution." (p. 251).

At this problem-solving stage the subject analyses the conditions of the task, tries to identify the required operations and establishes a plan in which these operations are correctly ordered.

During the execution of the action the subject is concerned with carrying out the plan created in the orienting stage. This implies that the orienting basis of an action determines in a large measure its nature and chances of success in problem solving. Therefore, in the theory of activity the importance of correct orientation in problem solving and during the process of "internalization" is often emphasized (Luria, 1966; Galperin, 1969; Leontiev, 1977).

3.2.2.2. *Luria's Interpretation of Defective Arithmetical Word Problem Solving after Frontal Lobe Damage*

Based on the just described approach, Luria has given an extensive description of defective arithmetical problem solving after frontal lobe damage (Luria, 1966; Luria & Tsvetkova, 1967; Luria 1973a).

This description starts from a general theory of thinking. In this theory, Luria (1973a) suggests that thinking arises when a subject is confronted with a problem for which he or she has no ready-made solution. In other words thinking arises in the presence of a task.

This task or problem often consists of a set of mutually subordinated conditions that have to be correctly encoded and carefully examined before any other cognitive step toward a solution can be undertaken. As a consequence, the solutor has to restrain impulsive responses and go through an orientation into the problem. This preliminary orientation comprises an investigation of the conditions of the problem, the analysis of its components, the recognition of essential features, and their correlation to one another. Such a preliminary investigation

> "is a vital and essential step in any concrete process of thought, without which no intellectual act could take place." (Luria 1973a, p. 327).

The next stage in the thinking process is

> "the selection of one from a number of possible alternatives and the creation of a general plan (scheme) for the performance of the task..." (Luria 1973a, p. 327).

This is the stage in which the solutor decides upon what strategy to use. Each task elicits different solving strategies, of which the solutor chooses the most adequate one.

According to Luria (1973a), preliminary orientation and choice of an appropriate strategy constitute

> "...the psychological essence of the processes of 'heuristics'..." (p. 328).

The next stage in thinking is the choice of appropriate methods and suitable operations for putting the general scheme of solution (strategy) into effect. This choice of procedures is part of the operational stage of problem solving.

Last but not least, Luria postulates that every intellectual act is concluded by a control stage, in which the obtained results are compared with the original conditions of the task. If the results do agree with these conditions, the intellectual act is concluded. If this is not the case, the search for an adequate strategy will continue until a solution in agreement with the conditions is found.

The Lurian conception of the thinking process can be schematically represented as follows:

STAGE	*SEARCH PRINCIPLE*
1. Preliminary orientation	} ⟶ Heuristic
2. Search for a strategy	
3. Search for procedures	⟶ Algorithm
4. Control	⟶ Feedback

Fig. 3-1: Stages and search principles in thinking, according to A.R. Luria (1973a).

Already earlier, Luria (1966) suggested that in complex cognitive tasks (e.g. arithmetical word problem solving) the disturbance characteristic of patients with frontal lobe damage is the absence of the stage of preliminary orientation. When given a problem, these patients usually omit investigation of the conditions that make problem solving possible and, as a rule, they do not analyse the givens essential for the solution. Without this preliminary activity, the frontal patient cannot formulate a plan or devise a strategy.

This absence of orienting activity after frontal lobe damage is found with different types of tasks; in constructive problems (Luria & Tsvetkova, 1964), in the understanding of printed stories and thematic pictures (Luria, 1966), in the perception of complex objects (Luria et al., 1966), and in arithmetical word problem solving (Luria & Tsvetkova, 1967).

According to Luria & Tsvetkova (1967) the following disturbances in the solution process of arithmetic word problems can be demonstrated:

1. The repetition of the givens and the final question of a word problem indicate whether the subject only encodes isolated fragments or whether he is able to retain the essential elements of the problem.
2. The analysis of the way in which a subject proceeds in solving a word problem shows whether the orienting base of his cognitive activity is unaffected or not. If the subject starts with an active analysis of the different conditions of the problem and tries to clarify the relations between these conditions, the orienting stage of the solution process is likely to be intact. Nevertheless, some subjects cannot find their way to the solution of the problem. The causes of their failure are then found in a defective *execution* of the necessary arithmetical operations. On the contrary, the orienting stage of the solution process is disturbed when a subject tries to solve the problem without any form of preliminary orientation and proceeds with arbitrary operations on some isolated fragment(s) of the problem.
3. The drawing of a general problem solution plan indicates that the subject uses a specific strategy and does not operate on isolated parts of the problem only.
4. The search for suitable operations can be selective and can remain within the limits established by the plan, or it can diverge from this plan and become influenced by casual external factors.
5. Errors in the solution process can be traced by examining the control stage in which, as was mentioned earlier, the subject compares the results with the initial givens and eventually corrects divergences.

Luria & Tsvetkova (1967) have thoroughly analyzed the solution process of arithmetic word problems in a few patients with different forms of brain damage. The first three of these patients suffered from tumors of the left posterior (parieto-occipital or parieto-temporal) parts of the brain, in all of the other cases the frontal parts of the brain were involved. The results of their investigation are summarized in Table 3-II. By means of a scrupulously careful analysis of think-aloud protocols, Luria & Tsvetkova were able to demonstrate several differences between left posterior and frontal patients in arithmetical word problem solving. Patients with posterior lesions failed to repeat word problems correctly, because they were not able to understand the meaning of logical grammatical propositions such as "less than", "longer than", "in order to", "some", and "how many times". On the other hand, patients with frontal lesions had few problems with the repetition of simple arithmetic word problems, whereas the repetition of complex problems was characterized by perseverations and simplifications. Moreover, the givens were often contaminated by givens of preceding problems.

The way in which the subjects of both groups tackled word problems also differed. The posterior group actively tried to grasp the meaning of the givens and the relation between different data, whereas frontal patients immediately began to solve some isolated fragment of the problem, without analyzing other givens further. In other words, the stage of orientation was still present in the first group, but largely absent in the second.

When they succeeded in surmounting their logical grammatical problems (often with some external cueing), patients with posterior lesions were even able to develop a plan or solution scheme, whereas frontal patients were not.

Disturbances in the execution of the plan also differed in both groups. Patients with left posterior lesions were not capable to recode linguistic propositions into mathematical operations. Such transformations were unproblematic for frontal patients, but the obtained operations did not fit into a plan and had a casual and fragmented character.

Finally, patients with posterior lesions were constantly aware of their difficulties and their errors, whereas frontal patients never noticed the contradictions and discrepancies they were subjected to (see Table 3-II).

Luria & Tsvetkova (1967) also discovered essential differences within the group of patients with frontal lobe damage. The solution process of patients with "frontobasal" (basomedial) lesions was seen against a background of disinhibited and impulsive behavior (as described under 2.1.2). As a consequence, these patients were perfectly capable of solving simple arithmetic word problems in which the resolution algorithm was unambiguously determined by the givens. When confronted with more complex problems, however, they omitted the stage of orientation, simplified the givens, responded to only a fragment of the problem or replaced complex reasoning by a stereotyped response, that had already been given. If another person, e.g. the experimentator, intervened and slowed down the uncontrolled way of responding by stimulating the repetition of the givens and the decomposition of the problem into its different parts, the patient was able to find his way to a solution.

The solution process of patients with "posterofrontal" (dorsolateral) lesions, on the contrary, was hampered by a lack of activity and a tendency to stick to once appeared stereotypes. Although the intentional and selective character of the solution process (with some help from the experimentator) was preserved, the presence of stereotypes still made solving difficult. This tendency was particularly evident when a repetition of the problem was

Table 3-II: Characteristics of the different stages in arithmetical word problem solving of patients with left posterior and frontal brain tumors (adapted from Luria & Tsvetkova, 1967).

TYPE OF LESION	left parieto-occipital or parieto-temporal lesion (3 cases)	left frontobasal lesion (3 cases)	left postero-frontal lesion (7 cases)	massive bilateral frontal lesion (1 case)	frontal subcortical lesion (1 case)	atypical frontal lesion (*) (2 cases)
1. Repetition of problem story	Disturbed: poor comprehension of logical grammatical structures.	Frequent perseverations, contaminations and simplifications. Final question is often omitted.	Correct in simple, but fragmented in complex problems. (presence of inert stereotypes).	Generally correct, but does not elicit solution process.	Often incomplete and simplified. Perseverations and contaminations.	Disturbed by simplifications and echolalic repetitions.
2. Orientation	Intact	Absent	Absent	Absent	Absent	Absent
3. Planning	Intact	Absent	Absent	Absent	Absent	Absent
4. Execution	Disturbed: faulty transformation of linguistic information into arithmetical operations and deficient execution of operations.	Operations are executed correctly, but do not fit in a plan.	Operations are executed correctly, but do not fit in a plan.	Operations are executed correctly, but do not fit in a plan.	Operations are executed correctly, but do not fit in a plan.	Operations are executed correctly, but do not fit in a plan.
5. Control	Intact	Absent	Possible with some external cueing	Absent	Absent	Absent

(*) Atypical frontal lesion: lesion to the frontal lobes with very mild or even absent disturbances of higher cortical functions. According to Luria & Tsvetkova (1967) the analysis of the disturbance of intellectual activity of patients with such lesions is particularly interesting because the changes in cognitive functioning often represent the only sign on which a topical diagnosis can be based.

demanded; the elements of the final question were often confounded with elements of the givens.

The characteristic lack of activity was even more pronounced in the case of massive bilateral lesion of the frontal lobe. This type of lesion led to a complete desintegration of every form of systematic cognitive activity. Even systematic instruction and cueing from the experimentator had no effect.

Similarly, the patient with a left subcortical frontal and both the patients with an atypical frontal lesion were characterized by a total

> "desintegration of the selective system of intellectual operations" (Luria & Tsvetkova, 1967, p. 160, p. 173).

As a result, these patients suffered from a deficiency in the stages of preliminary analysis, elaboration and execution of a plan and in control over the final result.

3.2.2.3. Evaluation

Luria & Tsvetkova's (1967) monograph is undoubtedly the most far-reaching effort to apply a general theory on the functional role of the frontal lobes in cognition to the issue of arithmetical word problem solving. It is the only study in which the question of arithmetical word problem solving after frontal and left posterior brain damage is investigated so specifically and comprehensively. Several other authors (Lhermitte et al., 1972; Christensen, 1975) have later confirmed the findings of Luria & Tsvetkova. Nonetheless, for several reasons the conclusions of all these descriptive single-case studies of arithmetical word problem solving after frontal lobe damage, including Luria & Tsvetkova's study, must be viewed with caution.

Firstly, the value of the descriptions of the frontal patients' word problem solving behavior, although clinically accurate, remains limited. Variations in the location and the type of lesion, in the educational and IQ-levels of the subjects and in the complexity of the word problems used for investigation reduce the reliability of the conclusions drawn in the various studies. Secondly, the exclusive reliance on single-case observations raises a transfer problem. The question of how the conclusions of such studies can be generalized to patients with the same type of brain lesion remains unanswered. Thirdly, the concepts used to explain defective arithmetical word problem solving are often ambiguous. Several objections can, for example, be made to the distinction between orientation and execution. The theoretical status of both concepts is ambiguous. On the one hand, orientation and execution are considered as stages in the development of mental actions (Galperin, 1969), on the other hand they are frequently described as overt problem solving behavior (Luria, 1966; Luria & Tsvetkova, 1967). Furthermore, orientation is sometimes defined as mental or motor activity preceding the execution of other mental or motor acts (Luria, 1973a), whereas in other cases it is considered as a specific function of actions (Galperin, cited in van Parreren, 1983, p. 9). Moreover, the operationalization of the concepts of orientation and execution is arduous, because the distinction between both concepts can only be made a-posteriori, after careful observation and analysis of an action (Galperin, 1969; van Parreren, 1979, 1983). Even the assessment of the emotional expression of the subject should indicate whether the action he performs has an executive or an orienting function (van Parreren, 1983). Finally, the concept of orientation or orienting activity is very general and includes various sub-activities such as the preliminary analysis of the conditions of a task (Luria, 1969), the recognition of its essential features (Luria, 1973a), the comparison of these features (Luria, 1973a), the formation of a program of action (Luria, 1969), and the choice between a number of equiprobable alternatives (Luria, 1973b). These considerations lead to the conclusion that, although useful for a global analysis of arithmetical problem solving and probably the best available to Luria and his collegues at that time, the notions of orientation and execution are difficult to characterize as well as to operationalize.

3.2.3. *The Cognitive Approach*

A third view on the issue of problem solving and mathematical ability is the cognitive approach.

The emergence of cognitive psychology in the sixties coincides with a revived interest in thinking and problem solving. Different explanations can be invoked for the renewed attention to both topics, but the most obvious ones are the limited explanatory power of associationist and neo-associationist theories (see section 3.1.) when confronted with complex behavior such as language and problem solving (Chomsky, 1959; Miller et al, 1960) and the advent of the programmable digital computer (Neisser, 1976). The computer not only provides a powerful tool to set up experiments and to analyze data thoroughly, but the activities of the machine itself may be compared to human cognitive processes. Computers open up the possibility to simulate some of these processes and to discover how they function.

Cognitive psychologists introduce the idea that mental processes can be studied without being directly observed. These processes occur internally, but can be indirectly inferred from behavior. In mental chronometry for example, the aim of the additive factor method (Sternberg, 1969) is to study the process that intervenes between S and R. This process is supposed to consist of a number of stages which perform different functions. These functions are called upon by different aspects of the stimulus and/or the response processing. Reaction times of subjects are influenced when those aspects are appropriately varied. In doing so, the experimenter can make inferences about the specificity of the different stages.

Although cognitive psychology has not succeeded in producing a comprehensive theory of human problem solving, a number of models and hypotheses about different aspects of the problem solving process have been developed.

3.2.3.1. The Concept of Problem Space

One of the major theoretical contributions of computer simulation is the idea of problem space. The problem space refers to the problem solver's representation of four elements:

- the initial state, in which the starting situation of the problem is represented
- the goal state, in which the final situation is represented
- the intermediate states, consisting of situations on the way to the goal
- the operators, or the actions that will achieve the goal.

Problem solving is conceived as a search through this problem space (Newell & Simon, 1972; Newell, 1980), in which the solver has to discover a sequence of state changes that will lead from the initial state to the goal state.

A good illustration of the concept of problem space is the eight-tile puzzle (Anderson, 1985). This puzzle consists of eight numbered movable tiles, arranged in a 3x3 frame (Figure 3-2). One compartment in the frame is always empty, allowing the solver to move an adjacent tile into it, creating a new empty space for the next move, and so on. The goal is to achieve a given configuration of tiles, starting off from a different configuration.

A problem could for example be to transform

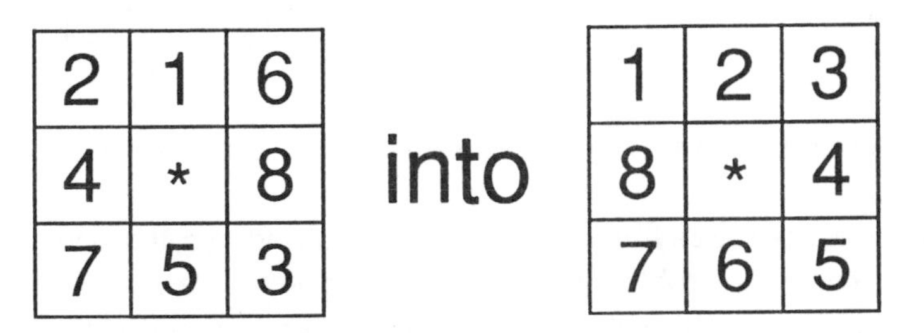

Fig. 3-2: Initial and goal states of an eight-tile puzzle.

The first configuration is the initial state, the second represents the goal state. Figure 3-3 reproduces the attempt of Anderson to solve the problem. His solution involves 26 moves, each move being an operator.

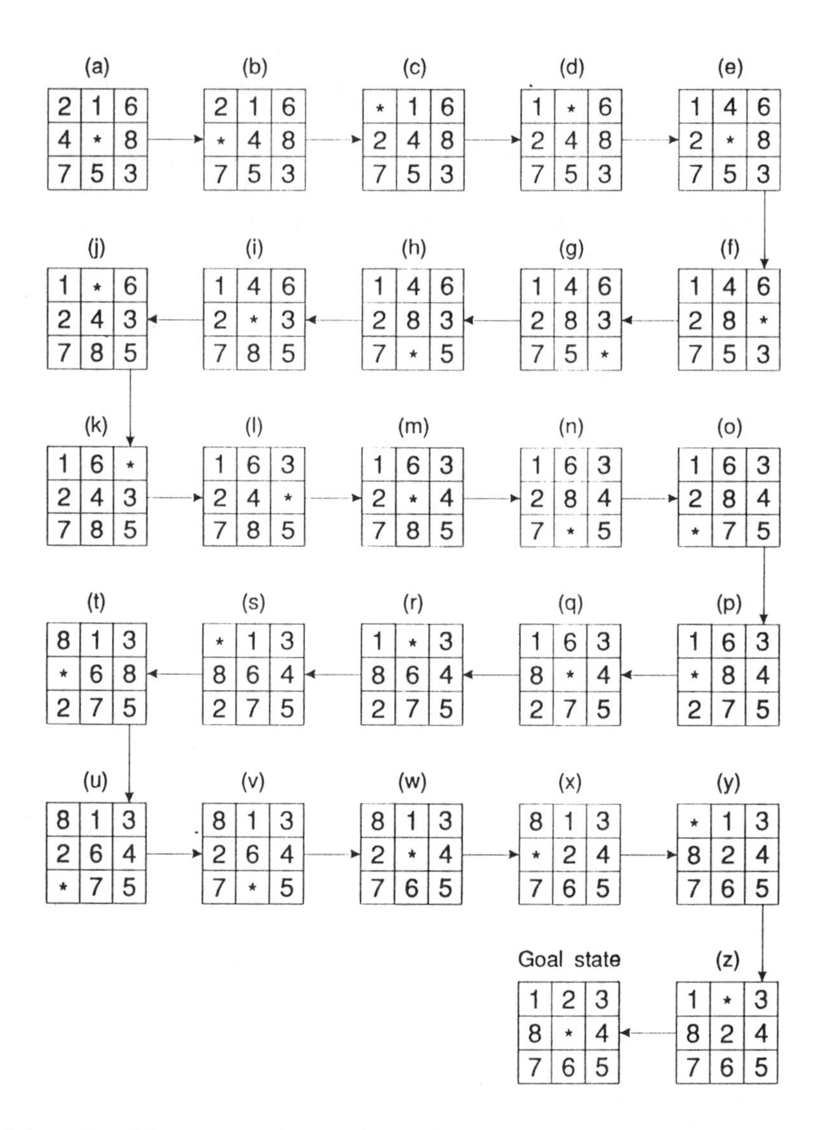

Fig. 3-3: Possible sequence of moves for solving an eight-tile puzzle (Anderson, 1985).

The sequence of operators illustrated in Figure 3-3 is longer than necessary. It is possible to solve the problem through a shorter sequence of moves.

The problem space is the set of all the states and operators that the problem solver is aware of (Ernst & Newell, 1969; Newell, 1980). Simon (1978) points out that the basic problem space – the problem space generated by a perfect problem solver – must be distinguished from a particular person's problem space. The latter problem space can be incomplete or contain errors. These errors can occur at two different stages in the problem solving process. In the first stage the problem solver translates the problem into an internal problem representation, that includes the initial state, the goal state and the "legal" or allowed operators. In the second stage the solver acts upon his problem representation by problem solving techniques (e.g. operators) in search of a solution. Until recently, very little attention has been paid to the errors that can occur in the translation process, although it is a vital step for correct problem solving (Mayer, 1983). Most efforts were, and still are, dedicated to the search strategies utilized to find a way from the initial to the goal state (see Anderson, 1985).

3.2.3.2. General Problem Solving Methods

It is obvious that problem solvers do not conduct an exhaustive search through the complete set of possible space-states in order to find the most efficient operators. Much of the research in problem solving deals with the identification of the principles that guide the search of the human problem solver through the problem space.

In the broad group of principles that have been discovered, it is usual to distinguish algorithms from heuristics.

Algorithms are solving procedures guaranteed to result in the solution of the problem at hand. They consist of a set of well-defined rules that can be applied to a variety of problems in the same category. An example of an algorithm in mathematical problem solving is the procedure of multiplication.

Heuristics, on the contrary, are less understood, but there can be no doubt that they play an important role in most problem solving tasks. They can be defined as a trick, a rule of thumb, a simplification or any other procedure that drastically limits the search for solution in the problem space. Johnson-Laird & Wason (1977) describe the difference between algorithms and heuristics as follows:

> " ... an algorithm is a 'mechanical' procedure that always yields the answer to a problem in a finite number of steps, whereas a heuristic procedure is one which offers a useful shortcut but which does not guarantee a solution." (p. 19).

This description indicates two fundamental differences between algorithms and heuristics; algorithms lead to a correct solution whereas heuristics not always do so. On the other hand, however, heuristics often yield a solution for the problem more rapidly than algorithms do.

It is impossible to enumerate all the heuristics that have been discovered in the solving of such diverse problems as analogies (Sternberg, 1977), crypt-arithmetic (Newell & Simon, 1972), the tower of Hanoi (Ernst & Newell, 1969), hobbits and orcs (Greeno, 1974), cheap necklace problems (Wickelgren, 1974) and many others.

The most significant heuristics are: hill climbing, in which the solver continually tries to reduce the difference between the initial state and the goal state; working backward, involving the reduction of the number of paths by working backward from the goal state towards the initial state; and solving by analogy, in which the method of solution to one problem is used as a guide to find the solution for another problem.

An extremely general and powerful method of problem solving is means-ends analysis (Newell & Simon, 1972). Means-ends analysis can be summarized as follows:

> "Given a desired state of affairs and an existing state of affairs, the task of an adaptive organism is to find the difference between these two states and then to find the correlating process that will erase the difference." (Simon 1969, p. 112).

In other words, in means-ends analysis, the problem solver tries to split up the problem into a set of differences and searches for operators that allow him to eliminate those differences. In Figure 3-4 two flowcharts illustrate the different stages of means-ends analysis.

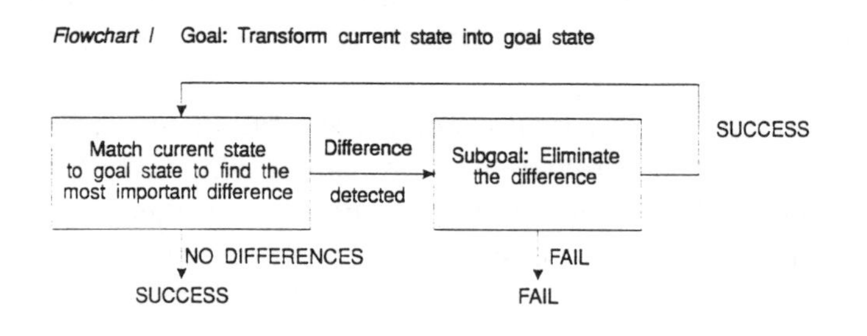

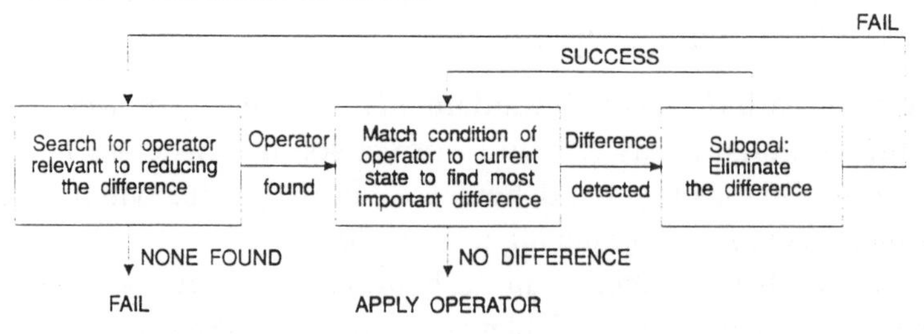

Fig. 3-4: Means-ends system of heuristic.

In the field of mathematical problem solving the usefulness of heuristics has been repeatedly quoted. Polya (1945) has explicitly recommended that attention should be paid to heuristic methods in the solution of mathematics problems. Ernst & Newell (1969) have discussed the application of means-ends analysis to algebra and calculus problems. Wickelgren (1974) has advocated the use of heuristics such as hill climbing, subgoaling and working backward in algebra word problems, geometry, trigonometry and calculus.

3.2.3.3. The Concept of Domain-specific Knowledge

In recent research a new condition for efficient problem solving has been emphasized; the availability of a well-organized body of domain-related knowledge. Evidence from a variety of sources supports this conception: data from experiments in developmental psychology (Chi, 1978; Heller & Greeno, 1978; Greeno, 1980a; Chi & Koeske, 1983), comparative studies of expert and novice problem solving (Larkin et al., 1980; Chi et al., 1981; Chi et al., 1982; Larkin, 1985) and process analyses of intelligence and aptitude (Pellegrino & Glaser, 1982).

The dominant role of domain-specific knowledge in problem solving had already been emphasized in the classic study of De Groot (1946) on the thought processes of chess players. De Groot ascertained that the fundamental difference between masters and mediocre chess players did not reside in the different strategies that both groups applied, but in the superior skills of memorization and recognition of chess patterns of the masters. Their superior achievements could be ascribed to their extensive knowledge of chess positions.

Chi (1978) compared a group of 10-year old expert chess players with a group of adult chess players that were newcomers in the field. The children scored considerably lower than the adults on a digit span memory test, but their memory for short presentations (10 seconds) of chess patterns was significantly better (Figure 3-5). Chi explains the better achievement of children in the chess problem by referring to their extensive knowledge of chess patterns.

Knowledge also plays an important role in the representation of the initial state of the problem and consequently in the understanding of the problem. Expert solvers of physics problems, for example, categorize these problems directly in terms of major physical principles, like the law of conservation of energy. Novice solvers' sort the same problems more in concrete object categories like "falling bodies" or "spring problems" (Chi et al., 1981).

There is increasing evidence that humans tend to categorize problems in order to understand them. Hinsley et al. (1977) have found that subjects tend to determine the type of problem they are going to deal with, almost immedi-

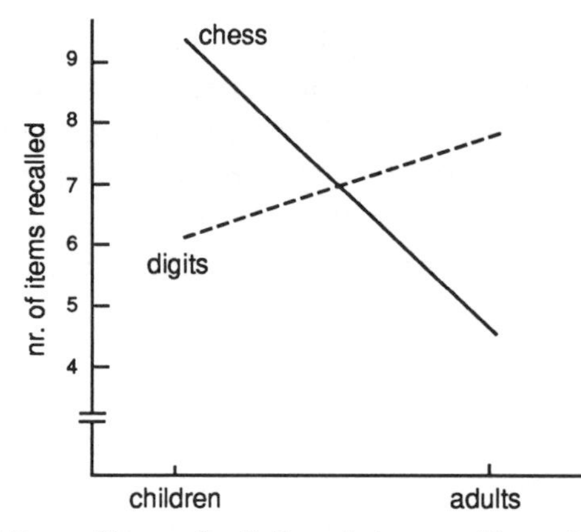

Fig. 3-5: Memory for digits and chess positions of children and adults (adapted from Chi, 1978).

ately after hearing the first couple of phrases. Robinson & Hayes (1978) submitted various problems to a group of subjects, asking them which information was important for a solution. Their subjects quickly categorized the problems and were able to distinguish relevant information from irrelevant information.

The knowledge of solution strategies is needed in order to establish and monitor plans for goals. Larkin et al. (1980) have studied the differences between the strategies used by experts and novices in solving physics word problems. Novices possess fragmented knowledge that compels them to ask "What do I do next?" at every stage. They tend to work backward from the goal to the givens, whereas experts have entire procedures in store that allow them to work forward from the givens to the goal.

Often the knowledge of appropriate algorithms is needed in order to carry out the procedures that lead to the solution of a problem. Addition, subtraction, division and multiplication are examples of algorithmic procedures that the solver of mathematical word problems has to master when he carries out computations.

In developmental psychology, the work of Groen & Parkman (1972) and Resnick (1976) proves that the amount of algorithmic knowledge differs in younger and older children. These authors were able to distinguish between different process models of children's addition and subtraction algorithms. Their work suggests that there is a developmental trend in which inefficient and less sophisticated models are used by younger children, whereas older children tend to use complex and more efficient ones.

All these studies clearly support the importance of the concept of domain-specific knowledge and indicate its close relation with skillful problem solving.

3.3. A Stage Analysis of Mathematical Word Problem Solving

The sharp distinction between performance based on general strategies and performance based on domain-specific knowledge is a theoretical one. Most modern theories of problem solving are based on the assumption that solving problems requires both general strategies and domain-specific knowledge (Greeno, 1980b; Larkin, 1980).

The better characterization of the performance of problem solvers in different kinds of tasks has also contributed to the erosion of the distinction between knowledge-based performance and general strategies.

An example of this improved characterization of problem solving is the stage analysis of mathematics word problems. The solution of such problems can be divided into a number of stages. Within these stages, several cognitive processes can be distinguished. Each process is characterized by the particular kind of knowledge (general and/or domain-specific) that it requires (Mayer, 1983; Mayer et al., 1984; Mayer, 1987). The "Astronaut Problem" of Mayer et al. (1984, p. 232), summarized in Figure 3-6, is a good illustration of the cognitive processes and the types of knowledge involved in the solution of an arithmetic word problem.

The next four sections explore the role of these four processes in mathematical word problem solving.

3.3.1. Encoding and Internal Representation

According to most theories of mathematical word problem solving, the solution of word problems starts with a complex text-processing activity aimed at properly encoding the problem. Starting from the verbal text, the solutor constructs an internal representation of the problem. (Mayer, 1983; Mayer et al., 1984; De Corte & Verschaffel, 1987). This representation is considered to be the result of a complex interaction of bottom-up and top-down analysis; that is, the processing of the different sentences of the problem as well as the activation of the subject's cognitive schemata contribute to the the construction of the problem representation. Word problem encoding consequently consists of two distinct processes:

1. a translation process, in which each sentence of the problem is transformed into a semantic memory representation, and
2. an integration process, in which the different sentences of the problem are put together in a coherent problem representation.

In the following section, the kinds of knowledge needed for each of these processes are examined.

1. The astronaut problem.

An astronaut requires 2.2 pounds of oxygen per day while in space. How many pounds in oxygen are needed for a team of 3 astronauts for 5 days in space?

2. Types of kwowledge required in solving the astronaut problem.

PROCESS	TYPE OF KNOWLEDGE	EXAMPLE FROM STORY
1. Translate	Linguistic and semantic knowledge	Variables are: (total oxygen), (oxygen per astronaut per day) (no. of days), (no. of astronauts).
2. Integrate	Schematic knowledge	"Time-rate" problem schema is total = rate x time. Total is (total oxygen). Rate is (Oxygen per astronaut per day).
3. Plan	Strategic knowledge	First figure out Time, (Time)=(no. of days) x (no. of astronauts). Then figure out Total, (Oxygen) = (Oxygen per astronaut per day) x time.
4. Execute	Algorithmic knowledge	Carry out procedure for 3 x 5, such as adding 5 + 5 + 5.

Fig. 3-6: Cognitive processes and types of knowledge required in mathematical word problem solving (based on Mayer et al., 1984).

3.3.1.1. Sentence Translation

The translation process of word problem sentences requires two kinds of knowledge: linguistic knowledge, i.e. knowing what the expressions in a word problem mean; and semantic knowledge, that is knowing what these expressions imply.

The "Astronaut Problem" of Figure 3-6 is a good illustration of the representational demands of a word problem.

When a solver tries to represent that problem, he has to know that "requires" is a verb and that "an astronaut" and "a team of astronauts" refer to the same variable; in other words, he needs linguistic knowledge. He also has to be aware that "day" is a measure of time consisting of 24 hours and that without "oxygen" an individual cannot live in space; to realize this, he needs semantic knowledge.

Greeno et al. (Heller & Greeno, 1978; Riley & Greeno, 1978; Greeno, 1980a) asked children to listen and then to repeat arithmetic word problems. The children could easily repeat the problems in which each sentence involved one variable, such as

> Joe had 3 marbles. Then Tom gave him 5 marbles. How many marbles does Joe have now?

When they had to repeat word problems including sentences dealing with a relation between variables, like

> Joe had 3 marbles. Tom has 5 more marbles than Joe. How many marbles does Tom have ?

the number of errors increased significantly. Apparently, sentences involving relational information are more difficult to represent in memory than one-variable sentences.

This is consistent with the findings of Mayer et al. (1984). In research with college students, they ascertained that assignment propositions of the type "The cost of candy is 1.70$" or "Total amount invested was 5.000$" were considerably easier to represent than relation propositions as "The length is 2,5 times the width" or "The rate in still water is 12 mph more than the rate in the current".

There is also evidence that minor changes in the representational demands of word problem sentences can have significant effects on the solution of arithmetic word problems. Maier & Burke (1967) for example, have demonstrated that representation 1 of the following horse business problem was poorly solved by their subjects (<40%), whereas the solution rate of representation 2 of the same problem was almost 100%.

> Representation 1
> A man bought a horse for 60$ and sold it for 70$. Then he bought *it back again* for 80$ and sold it for 90$. How much money did he make in the horse business ?
> (The right answer is 20$).
>
> Representation 2
> A man bought a white horse for 60$ and sold it for 70$. Then he bought *a black horse* for 80$ and sold it for 90$. How much did he make in the horse business ?
> (The right answer is 20$).
>
> (adapted from Maier & Burke, 1967, p. 305-309)

Computer simulations also suggest the important role of linguistic and semantic knowledge in mathematical word problem solving (Bobrow, 1968;

Hayes & Simon, 1974). The programs written for these simulations always contain some knowledge of the rules of English ("Pounds" is the plural of "pound", "Years younger than" means "less than", etc...) and some basic world-knowledge facts (1 foot equals 12 inches, distance equals speed times time, etc...). Without this type of information, the programs cannot solve the problems they have been written for.

More recently, several computer simulation models have been developed in an attempt to specify the internal processes and the kinds of knowledge involved in arithmetical word problem solving (Riley et al., 1983; Briars & Larkin, 1984; Dellarosa, 1986). In all these models the specific role of semantic knowledge in the representation of word problem sentences is emphasized.

3.3.1.2. Integration of Problem-information

In order to complete his representation of the problem, the solver must integrate the different parts of the problem into a coherent structure. Several studies have shown that problem-schema knowledge, or knowledge of problem types, is needed to connect the information of a problem into a comprehensible whole.

Hinsley et al. (1977) asked subjects to categorize algebra word problems from standard textbooks. They found high levels of agreement between subjects in the way they classified the problems. The subjects produced 18 different categories of problems such as average problems, ratio problems, area problems, etc... Apparently, the subjects knew the structure of at least 18 different problem types. After a compilation of all the word problems from the exercise sections of 10 major algebra textbooks and using these problems in a cued recall experiment, Mayer et al. (1984) found that high frequency word problems are easier to represent in memory than low frequency problems.

Several studies have evidenced that subjects use their problem-schemata in order to judge which information is relevant or irrelevant in a word problem (Hayes et al., 1977; Robinson & Hayes, 1978). Moreover, the recognition of a familiar problem-pattern provides quick access to relevant solution procedures. As already mentioned in section 3.2.3.3., Larkin et al. (1980) have proven that expert physics problem solvers are able to work problems from the bottom up, combining basic information in a sequence of steps towards the goal, whereas novices work backwards from the unknown problem solution to the given quantities. The experts apparently have a schema of the entire structure of the problem available, which allows them to take a direct path toward a solution. The same conclusion is drawn by Berger & Wilde (1987), after comparing expert and novice algebra word problem solvers.

The role of problem-schemata in problem representation has also been emphasized in computer simulation studies (Larkin et al., 1980). The ISAAC simulation program (Novak, 1977), for example, solves statics problems stated in words. The program generates a comprehensive representation of the problem situation, from which it directly derives appropriate equations for the solution of the problem.

3.3.2. *The Search in the Problem Space*

After having correctly encoded the information of a word problem, the solver has to search the problem space in memory for a solution. Within this search stage, two processes are important: planning and calculation. Planning presupposes strategic knowledge, whereas for calculation the solver draws on algorithmic knowledge.

Generally speaking, strategic knowledge refers to the solver's knowledge on how to establish and monitor plans for goals (Mayer et al., 1984; Mayer, 1987). In other words, the solver needs to know which operators to apply and when to apply them.

In the field of mathematical problem solving, several authors have proposed different strategies to help the solver find his way to a solution. A well-known attempt is that of Polya (1945), who suggests several general strategies for mathematical problem solving like restating the problem, looking for related problems and solving a part of the problem. The strategy most investigated is undoubtedly means-ends analysis. In physics word problem solving this strategy seems to characterize "novice" performance, but is not used by "expert" solvers (Larkin et al., 1980). Recent research has also shown that means-ends analysis is not always the most effective strategy in mathematical problem solving. It may interfere with the awareness of the structure of the entire problem (Owen & Sweller, 1985).

The strategy used by a problem solver is often influenced by the format of the problem. In other words, different problem representations may lead to qualitatively different solution strategies. Mayer et al. (1984), for example, have found that subjects use different strategies when they solve algebra equations and when they solve isomorphic problems stated in words.

Successful planning depends on two conditions (Anderson, 1983). Firstly, the solver must be able to spin forth goals, without acting upon them. Secondly, working memory has to be reliable enough to maintain the goal structure for operation.

Having decomposed the word problem, the solver needs to identify which operations are necessary to solve the different subgoals. The subgoals and operations that belong to the solution strategy of the following word problem, are listed below:

Floor tiles are sold in squares 30 cm on each side and weigh 10 g each. How much would it cost to tile a rectangular room 7.2 m long and 5.4 m wide if the tiles cost $0.72 each ?

Subgoals	Operations
1. Find the area of a tile	Multiply
2. Find the area of the room	Multiply
3. Find the number of needed tiles	Divide
4. Find the cost of needed tiles	Multiply

(adapted from Mayer, 1987)

Once the problem is represented and a solution strategy is devised, the problem solver has to carry out the necessary computations.

In order to carry out these computations, he needs basic algorithm knowledge, or knowledge of how to add, subtract, multiply, and so on. Algorithmic knowledge is the kind of knowledge that is maximally emphasized in mathematics instruction, whereas less attention is paid to other forms of mathematical knowledge like schematic or strategic knowledge (Simon, 1980, Mayer, 1986).

3.3.3. *Evaluation*

The preceding paragraphs illustrate that through the use of a cognitive approach, considerable progress has been made in revealing the cognitive processes and the knowledge-components involved in mathematical word problem solving. The described way of analyzing word problem solving is also consistent with developments in the information processing approach to cognition (Mayer, 1983).

The application of a cognitive approach to the issue of arithmetical word problem solving after brain damage should allow to overcome some of the drawbacks of a (possible) psychometric approach and of Luria's explanation. Moreover, it may throw a new light on the issue of word problem solving after frontal lobe damage.

The cognitive approach does not only focus on the abilities (e.g. perceptual speed or verbal comprehension)) that constitute task performance, like the psychometric approach does, but it is also interested in identifying the processes that underlie skillful task execution. The comparative investigation of these processes in brain-damaged and healthy subjects may reveal which stages in the information processing chain are specifically affected by the injury (Stokx & Gaillard, 1986). For brain damaged subjects such an investigation can also shed more light upon the exact locus of the residual effects of the injury. Moreover, cognitive information processing models allow predictions to be made about high-level cognitive impairments in brain-damaged subjects (Shallice, 1982).

In comparison to the ambiguous concepts of orientation and execution of Luria's description, the cognitive approach to mathematical problem solving offers some concepts with solid empirical support, derived from various sources such as computer simulation and novice-expert comparison. In addition, this approach requires that the hypothesis and conclusions about disabilities in mathematical word problem solving be stated in domain-specific or task specific terms, avoiding general explanations of word problem solving deficiencies.

Finally, an analysis in terms of cognitive processes may reveal how these processes can be manipulated in order to enhance recovery in frontal lobe patients.

3.4. Methodological Issues

3.4.1. Aims of the Investigation

The aim of the present study is to focus on the cognitive approach to mathematical word problem solving in order to investigate why patients with frontal lobe lesions fail to solve arithmetic word problems.

The application of this approach to arithmetical word problem solving may offer some new perspectives.

As explained above, the cognitive approach has resulted in a considerable amount of empirical evidence concerning the specific processes underlying mathematical word problem solving. Our analysis will be aimed at these specific processes, avoiding broad explanatory concepts of defective arithmetical word problem solving after frontal lobe damage. The cognitive analysis of mathematical word problem solving in stages may enable us to indicate where and how word problem solving difficulties arise for patients with frontal lobe lesions.

Moreover, this approach can be helpful in generating and applying rehabilitation measures pointed at particular stages in the arithmetical word problem solution process.

However, the scope of our study is limited.

Its main restriction concerns the cognitive processes that will be investigated. The present study will focus on the processes of sentence translation and information-integration in arithmetical word problem solving. Both these processes are a part of the *encoding* stage of word problem solving (Mayer, 1983; Mayer et al., 1984; Berger & Wilde, 1987). In information processing terms, encoding precedes the stage of problem space search, in which a solution strategy or a solution plan is devised and arithmetical computations are executed. Strategic knowledge and planning have already been extensively investigated in patients with frontal lobe damage, and the lack of planning capacities has been repeatedly indicated as one of the main symptoms characterizing the frontal patient's behavior (Luria, 1966; Lhermitte et al., 1972; Lezak, 1982; Shallice, 1982). On the other hand, there is significant empirical evidence that patients with frontal lobe lesions still have enough procedural knowledge available in order to carry out arithmetical calculations such as additions and subtractions without particular difficulty (Christensen, 1975; Walsh, 1978). As mentioned in section 2.3.1., these arithmetical calculation disorders are characteristic of patients with posterior brain lesions. Given this, the detection of encoding impairments could yield a more complete description of the arithmetical problem solving impairments of patients with frontal lobe damage. Moreover, it could change our view on defective arithmetical word problem solving after frontal lobe lesions as a planning problem only.

A second restriction concerns the distinction between two fundamental requirements for successful problem solving: an information processing system with intact structural characteristics (e.g. an adequate speed of information processing) and the availability of domain-specific knowledge (e.g. the knowledge of solving strategies). As already mentioned, the focus of the present study will distinctly be on two cognitive processes required for skillful arithmetical word problem solving, namely sentence translation and integration of problem-information. Most of the tasks used to investigate these processes will require both the functioning of information processing structures as well as a sizable amount of domain-specific knowledge. Therefore, a sharp distinction between both aspects of problem solving ability will not be drawn in the present study.

3.4.2. Method of Investigation

In order to study sentence translation and information-integration in patients with frontal lobe lesions three separate experiments have been carried out.

The characteristics of frontal impairment in arithmetical word problem solving have been assessed by comparing the performances of a group of subjects with frontal lesions to those of a group of healthy controls *and* to those of a group of patients with left posterior lesions.

In Experiment 1 (Chapter IV) the translation of various types of word problem sentences to an internal representation was investigated. This translation process was examined in patient and control groups with a recognition task and a sentence-picture matching task. In the recognition task, subjects were asked to read, but not to solve, several arithmetic word problems. After each reading period they had to identify several sentences of different logical grammatical complexity. The same types of sentences were read aloud by the experimenter in the sentence-picture matching task, after which the subjects had to compare the sentences with various pictures. In addition, the relationship between sentence representation and arithmetical word problem solving was investigated. For this, every subject was asked to solve a series of arithmetic word problems of increasing complexity. The predictive value of sentence recognition for word problem solving performance was tested by means of a multiple regression analysis.

In Experiment 2 (Chapter V) the recognition of problem-schemata or word problem patterns was examined in the same three groups. Every subject was presented with a deck of 24 filing cards with an arithmetic word problem printed on each card. The subjects were asked to sort the cards into groups, based on whatever criteria they thought were relevant. The results of this word problem sorting task were analysed quantitatively as well as qualitatively. The implications of different kinds of sorting behavior for arithmetical word problem solving are discussed.

The purpose of Experiment 3 (Chapter VI) was to examine whether a cueing procedure aimed at improving the two already investigated text-processing skills, i.e. sentence translation and integration of information into a problem-schema, could ameliorate arithmetical word problem solving performance differentially. All subjects participating in the study were asked to solve two different series of analogous arithmetic word problems. The first series consisted of word problems that had to be solved directly after they had been read aloud by the subject. The second series was analogous to the first, but after reading each problem the subject had to answer several questions dealing with sentence translation and integration of problem-information. Verbal feedback on the correctness of the answers was given after each question, whereupon the subject was asked to solve the problem at hand. This procedure was repeated after a short time-interval in order to see whether the possible benefits of the cueing procedure remained stable over time.

CHAPTER IV

The Internal Representation of Arithmetical Word Problem Sentences: Frontal and Posterior-Injured Patients Compared*

4.1. Introduction

Frontal lobe damage can cause several impairments in cognitive functioning. One of the most subtle of these impairments is the inability to solve arithmetic word problems (Luria, 1966; Luria & Tsvetkova, 1967; Lhermitte et al., 1972; Christensen, 1975; Walsh, 1978). Although frontal patients still possess some basic mathematical skills, the solution of arithmetic word problems is particularly difficult for them. The preserved abilities include the capacity to execute elementary arithmetical operations such as addition and subtraction (Christensen, 1975; Walsh, 1978), the ability to cope with the visual-spatial organisation of number elements (Luria & Tsvetkova, 1964; Hécaen, 1969) and usually the conservation of the formal structures of language such as phonetics and syntactics (Luria & Tsvetkova, 1967; Novoa & Ardila, 1987).

Several explanations have been put forward to account for the incapacity of patients with frontal lobe damage to generate correct solutions in arithmetical word problem solving. According to Luria & Tsvetkova, who presented a

* This chapter is a slightly modified version of L. Fasotti, P.A.T.M. Eling, J.J.C.B. Bremer (1991). The internal representation of arithmetical word problem sentences: frontal and posterior-injured compared. Accepted for publication in Brain and Cognition.

detailed analysis of the problem in their French monograph (1967), patients with frontal lobe lesions primarily omit any form of preliminary orientation into arithmetic word problems. They are unable to inhibit the first tendency aroused by the problem and they start to perform mathematical operations with the fragment that has caught their attention. This kind of impulsivity impedes the formulation of a comprehensive strategy for the solution of the problem. A very similar explanation of defective mathematical word problem solving after frontal lobe damage is given by Lhermitte et al. (1972).

On a more descriptive level, Christensen (1975) states that the patient with frontal lobe damage only grasps one particular fragment of word problems. Without making any plans he starts to carry out arbitrary arithmetical operations with this fragment. These impulsive and fragmentary arithmetical operations are often unconnected to the final goal.

Barbizet (1970), investigating memory impairments with arithmetical word problem tasks (e.g. What is the length of one quarter of the Eiffel Tower?), explains the problems of frontal lobe patients in terms of a particular memory deficit. This memory deficit would affect the correct way to proceed in the solution process rather than the information necessary to solve the problem. If the examiner helps by breaking down the problem into different stages, asking for a response at each step, a correct answer can be achieved by these patients.

In contrast with frontal patients, patients with lesions in the left parieto-temporo-occipital region (posterior lesions) repeatedly attempt to analyze the givens of arithmetic word problems and they often try very hard to discover the information required for the formulation of a solution-strategy (Luria & Tsvetkova, 1967; Christensen, 1975). These attempts are, however, severely hampered by the poor understanding of complex sentences of a comparative or attributive type, e.g. 'the father is older than the son'...'John gives 12 marbles to Peter'.

Moreover, these patients are unable to transform such complex logical grammatical constructions into mathematical operations (Luria & Tsvetkova, 1967).

To shed more light on broad concepts such as "preliminary orientation" (Luria, 1966) or "impulsive conclusions" (Christensen, 1975), it could be useful to investigate the issue of arithmetical word problem solving from an information processing point of view. Such an approach would enable us to point out at which stage(s) of the information processing chain solution problems occur. The same approach could also be used to investigate Barbizet's (1970) claim that frontal damage should impair the application of procedures as such. Barbizet's explanation (pre)supposes that frontal patients possess the information necessary to solve a word problem, but that they act as if they had forgotten the correct way to proceed (Walsh 1978). An information processing approach could be helpful to investigate this assumption that frontal patients should dispose of sufficient information, and eventually refute Barbizet's hypothesis on the inadequate recall of procedures.

Although there is no generally accepted information processing model for word problem solving, most theories of mathematical word problem solving (Larkin et al., 1980; Mayer et al., 1984; Mayer, 1985; De Corte & Verschaffel, 1987) agree that the first stage in the solving process of word problems consists of a text-processing activity. Starting from the verbal text, the solutor constructs a global internal representation of the problem. This problem representation stage can be broken down into two main phases:

1. translation or correct encoding and comprehension of each individual sentence of the problem and,
2. integration i.e. putting the sentences together in a coherent and integrated representation of the problem (Mayer et al., 1984; Mayer, 1987).

The present study is mainly devoted to the first of these stages (translation) and to the relation between translation skills and the ability to solve arithmetic word problems. Both these aspects will be investigated with three groups: a group of patients with unilateral and bilateral frontal lobe damage, a group of patients with left posterior lesions and a group of healthy controls.

Two types of tasks have been used to investigate translation processes. The first type includes studies involving the use of short-term memory as a means for efficient translation. Greeno (1980a) for example, asked children to listen to arithmetic word problems and then to repeat them. The results show that problems including sentences which deal with only one variable are quite easy to repeat, whereas many repetition errors are made with sentences involving a relation between two variables. Recall experiments (Mayer, 1982; Mayer et al., 1984) with adults have confirmed these results. Apparently, sentences expressing relations are more difficult to encode properly than one-variable sentences.

A second series of tasks used to test translation skills consists of a transformation of arithmetical problem sentences into equations (Soloway et al., 1982), pictures (Mayer, 1987) or material representations such as puppets or blocks (De Corte & Verschaffel, 1987). The results of these studies confirm the special difficulty that the translation of relational propositions presents.

Research involving brain-damaged subjects has shown that in spontaneous speech, patients with frontal lobe lesions tend to produce simple sentences rather than compound, complex ones (Kaczmarek, 1984). Moreover, their comprehension of the logical structures of language is limited (Novoa & Ardila, 1987). The previously mentioned study of Luria & Tsvetkova (1967) demonstrates that patients with left posterior lesions also poorly understand complex sentences. From this empirical evidence we can hypothesize that error rates in the translation of word problem sentences increase with growing grammatical logical sentence complexity.

Patients with frontal lobe damage differ considerably from healthy controls in their ability to handle complex grammatical sentences (Kaczmarek, 1984; Novoa & Ardila, 1987). From this we can predict that difficulties in the translation of word problem sentences are more obvious in patients with frontal lobe damage than they are in a group of healthy controls.

Within frontal groups, patients with left hemisphere damage perform worse than patients with right hemisphere damage only when verbal memory is called upon (Novoa & Ardila, 1987). As a consequence, in a verbal memory task patients with right frontal hemisphere lesions should perform better than patients with left and bilateral frontal damage.

The meaning of words expressing comparisons (more-less, younger-older), complex prepositions and conjunctions presents a particular difficulty to the patient with lesions to the left parieto-occipital or parieto-temporal regions. These difficulties have repeatedly been quoted as aphasic disturbances under the heading of "semantic aphasia" (Head, 1926; Luria, 1966; Kertesz, 1979). Therefore, we can expect the error rates in translation tasks to be higher in the group with left posterior lesions than in the groups with frontal lobe damage and the group of healthy subjects. This should be particularly evident in word problem sentences containing expressions of the attributive or comparative type.

Finally, if the aforementioned theories of mathematical word problem solving are valid, poor translation skills should involve reduced mathematical word problem solving ability.

4.2. Subjects

Forty patients and 10 controls participated in this study. Thirty patients had frontal lobe lesions (10 bilateral, 10 in the left and 10 in the right hemisphere) and 10 patients left posterior lesions.

The site of lesion was identified as follows. For 7 patients extensive CT-scan reports were found in the medical file. In these radiologists' reports the location of the lesion was described in detail. For the remaining 33 patients, of whom no such detailed reports were available, the site of lesion was directly assessed from the CT-scans. The assessment consisted in a mapping of the site and the extent of lesion on a standard lateral diagram of hemisphere, following a procedure described by Mazzocchi & Vignolo (1978).

A lesion was defined as frontal if it was limited to cortical-subcortical areas located anterior to the Rolandic fissure. Frontal patients whom a speech pathologist judged aphasic were excluded. In most cases aphasia was assessed with the Aachen Aphasia Test (Huber et al., 1984).

Left posterior lesions were situated entirely behind the Rolandic fissure. In each of the 10 patients the parietal lobe was involved, either in conjunction with the occipital lobe or with temporal areas. Patients were excluded when we noticed reading difficulties or a severe form of receptive aphasia, as measured by the Token Test (De Renzi & Vignolo, 1962). Moderate comprehension problems were detected in 3 of the subjects with left posterior lesions. As these problems barely hindered the comprehension of task instructions, these subjects were not excluded from the study.

The etiology of the brain-damaged group was tumoral in 15, traumatic in 12, and vascular in 13 patients.

The controls were pupils of a secondary level professional education course in hairstyling, garage mechanics and secretarial work. These subjects were chosen in order to match the IQ-levels of the frontal and left posterior groups. Luteijn & Van der Ploeg (1983) have gathered IQ-data for different levels of education in the Dutch population. Based on this study, the educational level of the control group was used to estimate its mean IQ-level at 104 (sd= 11). Details with respect to sex, age, level of verbal and performal intelligence of patient groups and controls are given in Table 4-I. Age limits for patients and controls were drawn at 18 and 60 years.

Table 4-I: Patient and control profiles.

		bilateral frontal (n=10)	left frontal (n=10)	right frontal (n=10)	left posterior (n=10)	controls (n=10)
sex	male	7	6	7	5	6
	female	3	4	3	5	4
age	m	37.8	42.2	28.9	46.8	33.3
	s.d.	12.6	9.7	9.2	8.7	6.4
WAIS-R PIQ	m	101	109	107	100	104
	s.d.	16	12	19	11	11
WAIS-R VIQ	m	98	105	110	79	104
	s.d.	15	14	16	15	11

4.3. Procedure

In a separate session and prior to the experimental tasks, an extensive battery of neuropsychological tests, mainly assessing frontal functioning, was given to all the patients. The results of these tests are listed in Table 4-II.

Table 4-II: Neuropsychological test scores for different patient groups.

	bilateral frontal (n=10)		left frontal (n=10)		right frontal (n=10)		left posterior (n=10)	
	m	s.d.	m	s.d.	m	s.d.	m	s.d.
1. Rivermead Behavioral Memory Test								
– profile score	56.3	8.3	60.2	7.7	64.5	6.7	50.8	8.2
– screening score	8.7	2.0	9.0	2.0	9.9	1.1	7.2	2.8
2. Dutch Version of Rey's 15-word Test								
– immediate recall decile	3.3	1.8	4.8	2.5	5.0	2.6	2.1	1.9
– delayed recall decile	2.7	2.9	4.6	3.2	6.2	2.2	3.1	3.2
3. Modified Card Sorting Test								
– numers of categories completed	4.1	2.2	4.3	2.1	4.2	2.2	5.6	.9
– numer of perseverative errors	5.0	5.5	4.0	4.3	3.8	4.5	3.1	2.1
4. Tower of Hanoï								
– mean number of correct items	11.3	2.4	11.8	1.3	12.3	.9	11.2	1.6
– mean number of moves	3.6	.9	3.0	2.1	2.6	1.1	4.5	1.7
5. Hooper Visual Organization Test								
– total score	20.1	3.3	24.5	3.1	22.9	3.8	20.5	4.7
6. Porteus Mazes Test								
– mean time	49.6	26.2	37.1	17.4	30.3	17.1	60.8	34.1
– mean number of errors	1.6	1.6	.5	.7	.6	.6	1.3	1.2

After these neuropsychological tests, three tasks were presented in the same order to all subjects. First, every subject was administered a recognition task with different types of arithmetical word problem sentences. Then, each subject performed a sentence-picture matching task with three types of arithmetical word problem sentences. Finally, patients and controls were asked to solve a series of arithmetic word problems of growing complexity.

Sentence recognition task

In the sentence recognition task patients and controls were asked to read, but not solve, a series of 11 arithmetic word problems, spending one minute on each problem. After each period of study, the text of the word problem was taken away and 8 sentences printed on cards were sequentially shown to the subject, who was asked to judge for each sentence whether or not it had occured in the word problem just presented. The yes-no responses were given orally. The order of presentation of the word problems and the recognition sentences was fixed.

The word problems contained 4 kinds of propositions:

1. *Assignment propositions.* These propositions gave a single numerical or indetermined value for a person, e.g. "John has 7 marbles" or "Martin has some stamps". Apart from an identical match, the recognition cards could display two kinds of mismatches: propositions with quantity specification errors ("John has 3 marbles") and propositions with person specification errors ("Martin has 7 marbles").

   In both types of mismatch, a variable of the original assignment proposition was replaced by the variable of another sentence of the same word problem.
2. *Relation propositions.* Relation propositions gave a single numerical or indetermined relationship between two persons. Examples included "Peter gives 5 marbles to John" or "Eric gives some stamps to Willy".
3. *Compare propositions* were propositions in which a variable was defined in terms of another variable, such as "Ben has 6 stamps more than Martin" or "Jack has some marbles less than Peter".

   Relation and compare propositions are more difficult to translate than assignment propositions (Mayer et al., 1984; De Corte & Verschaffel, 1985). In relation propositions, beside specification errors, mismatches could bear on subject inversions (e.g. "John gives Peter 5 marbles" instead of "Peter gives John 5 marbles"). In compare propositions logical grammatical complexity is increased by the presence of words expressing comparison (e.g. as many as/more/less, younger/older). In this kind of sentences mismatches were related to a wrong comparative expression (like "Mike has 3 apples less than Phil" instead of "Mike has 3 apples more than Phil") or to a subject inversion, changing the meaning of the compare proposition (e.g. "Phil has 3 apples more than Mike" instead of "Mike has 3 apples more than Phil").

Alternatives implying cognitive transformations such as the matching of "Mike has 3 apples more than Phil" with "Phil has 3 apples less than Mike" were not used.

4. *Question propositions.* These propositions involved a question concerning either (1) a single numerical value for some person, such as "How many marbles has John left?" (assignment questions), or (2) a relation question concerning a single numerical or indetermined value for more than one person, e.g. "How many stamps did Eric give to Jack?" or (3) a comparative question such as "How many apples does Eddy have to give away in order to possess as many as Ron does?".

Sentence-picture matching task

In the sentence-picture matching task 15 assignment, 7 relation and 6 compare propositions were presented to patients and controls.

Each proposition was read aloud by the experimenter, after which the subject was asked to compare the sentence to 5 successive pictures in A4 format. Each subject had to judge whether or not the pictures did represent the sentence correctly. Two out of the 5 matches were correct; these correct matches differed from each other only in the color of the depicted objects. The order of presentation was identical for every subject.

In assignment propositions, the same specification errors were possible as in the preceding sentence recognition experiment. An example of pictures presented after the assignment proposition "John has 5 marbles" is given in Figure 4-1. Complex propositions were again of the relational or comparative type. In relation propositions two kinds of mismatches could occur: specification errors and inversion errors. Inversion errors were made when a picture showed the action of a sentence in reverse (e.g. the sentence "John gives Eric two towels" was matched with a picture in which Eric gives two towels to John) and subjects answered "yes" but also when subjects answered "no" to correct match pictures.

In compare propositions, specification mismatches bore on the number of objects shown in the picture. Figure 4-2 shows the 2 match and the 3 mismatch pictures presented after the compare proposition "Suzan has more sweaters than Jane".

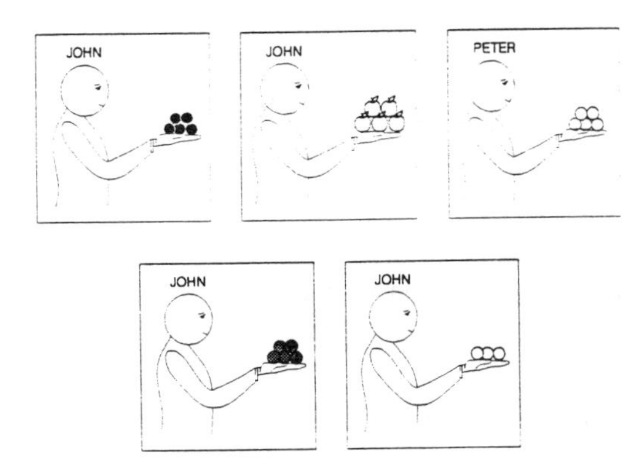

Fig. 4-1: Pictures presented after the proposition "John has 5 marbles".

Fig. 4-2: Pictures presented after the proposition "Suzan has more sweaters than Jane".

Table 4-III: Arithmetic word problems and number of arithmetical operations.

WORD PROBLEMS	NUMBER OF OPERATIONS
1. Carl has 3 pencils and John has 5 pencils. How many pencils do they have together?	1
2. A car shop has 60 spark plugs for a total worth of 180 $ in stock. What is the price of a sparking plug?	1
3. Paul has been jogging for 11 miles. Frank has been jogging 4 miles less than Paul. How many miles have they been jogging together?	2
4. Ellen has 5 apples. Karen has 4 times more apples. How many apples do they have together?	2
5. Company A has 20 employees. Company B has 3 employees less than company A. Company C has 5 employees more than company A. How many employees do the 3 companies have in total?	3
6. Grandmother A has 18 grandchildren. Grandmother B has 7 grandchildren less than grandmother A. Grandmother C has one-third of the number of grandmother A's grandchildren. How many grandchildren do the grandmothers have in total?	3
7. Peter has 10 marbles. His sister has 21 marbles more. Peter's brother has 5 marbles less than Peter and his sister together. How many marbles do the 3 children have together?	4
8. Peter is 12 years old. His brother is 5 years younger. His grandfather is 4 times older than Peter and his brother together. How old are the 3 together?	4
9. An has 22 $. Maud has 17 $ less. Sarah has 9 $ more than An and Maud together. How many $ does every girl have on average?	5
10. A shopkeeper has sold 11 bottles of milk. A second shopkeeper has sold 3 bottles less. A third shopkeeper has sold 4 bottles more than the first and second shopkeepers together. Together the 3 shopkeepers have 11 bottles of milk left. How many bottles did they buy together?	5
Total score	30

Word problem solving
Finally, every subject was asked to solve 10 arithmetic word problems of increasing complexity. Complexity was defined as the number of arithmetical operations required to solve the problem. The order of problem presentation was randomized and identical for all subjects. The 10 word problems and the number of arithmetical operations needed for each problem are presented in Table 4-III.

4.4. Results

Recognition task
In order to enhance reliability, several recognition items were excluded from each proposition and question category, namely those with a Cronbach's $\alpha < .65$. Table 4-IV illustrates the original as well as the residual number of recognition items and Cronbach's α for each type of proposition and question.

Due to its small number of residual items and to a low internal consistency, the category of compare questions was excluded from further analysis.

Figure 4-3 shows the proportion of recognition errors for experimental and control groups.

Table 4-IV: Original and residual number of items after deleting items with low inter-item consistency.

	assignment propositions (n=7)	relation propositions (n=6)	compare propositions (n=4)	assignment questions (n=5)	relation questions (n=3)	compare questions (n=3)
number of recogniti-on items	20	19	16	15	10	8
residual num-ber of recog-nition items	14	14	11	11	8	4
Cronbach's α	.75	.76	.73	.76	.68	.59

A two-way ANOVA with repeated measures on recognition errors for three types of propositions indicates a significant group ($F= 13.38$, $df= 4/45$, $p<.01$) and complexity ($F= 5.65$, $df= 2/90$, $p<.01$) effect, but no significant interaction between groups and complexity ($p=.43$).

Multiple t-test comparisons between the different types of propositions reveal significant differences between assignment and relation propositions (t=-3.04, p<.01), between assignment and compare propositions (t=-3.15, p<.01), but not between relation and compare propositions (p=.62). The results suggest that assignment propositions are considerably easier to translate into memory than relation and compare propositions, but that such a translation is equally difficult with relation and compare propositions. Although an examination of the error rates presented in Figure 4-3 could suggest that a difference between relation and compare propositions is present in the left posterior group, the comparison fails to reach significance (p=.18).

Further investigation of the significant group effect with multiple contrast analyses reveals that recognition error rates in the frontal and the left posterior groups are significantly higher than in the group of healthy controls (T=6.97, df=33.6, p<.01 and T=-6.15, df=9.5, p<.01 respectively). Frontal patients perform much better than patients with left posterior lesions (T=-3,32, df=11.3, p<.01).

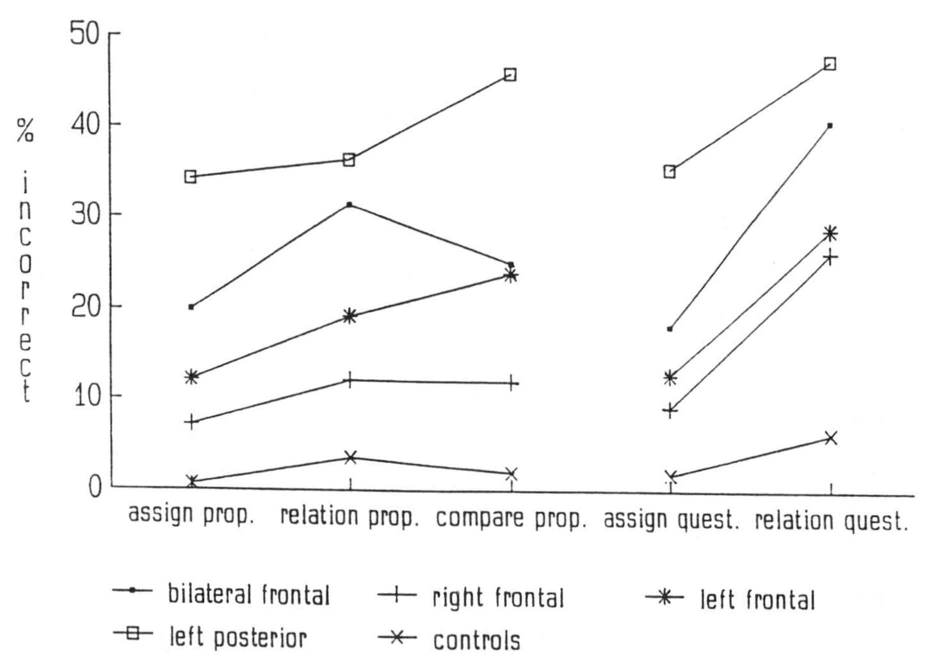

Fig. 4-3: **Mean percentages of recognition errors in 5 groups for different types of propositions and questions.**

A Newman-Keuls test applied to the error rates within the frontal group yields a significant difference between the right and bilateral frontal groups (p<.05) whereas the right-left frontal and left-bilateral frontal comparisons are short of statistical significance.

A two way ANOVA with repeated measures of question recognition error rates reveals a significant group (F=8.07, df=4/45, p<.01) and complexity effect (F=19.41, df=1/45, p<.01), whereas the group-complexity interaction effect does not attain significance (p=.53).

The significant complexity effect indicates that relation questions are more difficult to translate into short-term memory than assignment questions.

The groups effect was further analyzed with multiple contrasts. These contrasts reveal that error rates in the frontal and left posterior group are again significantly higher than in the healthy control group (T=6.41, df=30.4, p<.01 and T=-5.19, df=9.3, p<.01). Patients with left posterior lesions commit considerably more question recognition errors than frontal patients (T=-2.47, df=11.7, p<.05).

Within the frontal group, no appreciable differences are found among the three groups with different lesion localization.

Sentence-picture matching task

In comparing sentences with pictures, the healthy control subjects made no errors. The error percentages of the different patient groups for assignment, relation and compare propositions are illustrated in Figure 4-4.

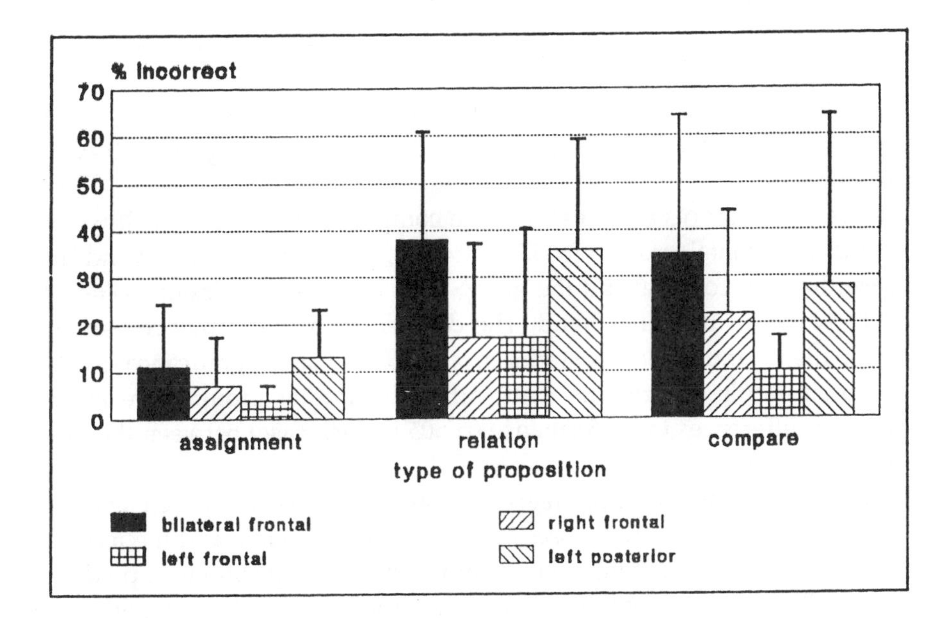

Fig. 4-4: Mean percentages (and sd's) of errors in 4 patient groups for three types of propositions in the sentence-picture matching task.

For patients, assignment propositions are easier to match with pictures than relation (t=-3.89, p<.01) and compare (t=-3.25, p<.01) sentences. There are no significant differences between relation and compare sentences (p=.61).

Despite large differences in means, no significant differences between patient groups were found, due to large within group variances.

Word problem solving

The results of the arithmetical word problem solving experiment are given in Table 4-V, in which the mean number (and sd) of omitted arithmetical sub-operations per group is given. For each non-identified step or its corresponding operation one error point was given. Errors in the execution of calculations were not counted. The highest possible error score over 10 word problems was 30 points (see Table 4-III).

Table 4-V: Mean error scores (and sd's) for experimental and control subjects in the arithmetical word problem solving task.

	bilateral frontal (n=10)	left frontal (n=10)	right frontal (n=10)	left posterior (n=10)	controls (n=10)
mean	7.4	5.7	2.9	9.2	1.9
s.d	4.45	5.68	3.18	8.44	1.20

The results of a one-way analysis of variance suggest that there are significant differences in problem solving skills between our five groups (F=3.43, df=4/45, p<.05).

Relevant differences are present in the comparisons between healthy controls and the frontal group (T=-3.76, df=29.8, p<.01) on the one hand and between healthy controls and the group with left posterior lesions (T=-2.70, df=9.4, p<.05) on the other hand. The differences in performance between frontal patients and the left posterior group do not reach significance (p=.20). A Newman-Keuls test reveals that within the frontal group there are no significant differences in performance (p<.05 in each case) between the three frontal subgroups.

Finally, a stepwise multiple regression analysis was performed with the word problem solving error scores as a dependent variable and the recognition error rates of assignment, relation and compare propositions as predictor variables. This analysis reveals that high error rates in the recognition of assignment propositions (R^2= .40, Beta=-.63, sign. T<.01) are the best predictor for poor arithmetical word problem solving ability. Next best predictor is the error rate in the recognition of compare propositions (R^2=.45, Beta=-.31, sign. T<.05), whereas the contribution of relation propositions to the explained variance only approaches conventional levels of significance (Beta=-.29, sign. T=.09).

The same stepwise procedure with the recognition error rates of assignment and relation questions as predictor variables reveals that the faulty recognition of relation questions (R^2=.26, Beta=-.36, sign. T<.01) is the best predictor for low levels of arithmetical word problem solving. Assignment questions (R^2=.33, Beta=-.29, sign. T<.05) also significantly predict high error scores in arithmetical word problem solving.

4.5. Discussion

The main purpose of this study was to test three major hypotheses about the cognitive representation of word problem sentences in normals, in patients with frontal lobe lesions and in patients with posterior left hemisphere pathology. In addition, the relation between sentence representation and word problem solving skills was investigated.

Our first hypothesis, concerning the influence of increasing grammatical logical complexity on recognition error rates, was only partially confirmed. On the one hand, the translation of assignment propositions into internal representations is indeed much easier than the translation of relation or compare propositions. On the other hand, there are no significant differences in the representation of relation and compare propositions. Apparently, the left posterior group is the only group in which recognition error rates are higher in compare than in relation propositions. This difference, however, is not statistically significant.

The assumption that compare propositions might be more difficult to represent than relation propositions was mainly based on research with young children (Greeno, 1980a; Riley et al., 1983). In these studies, erroneous word problem solving resulted from a deficient knowledge of complex logical grammatical expressions. Apparently, once these constructs are mastered (e.g. in adulthood) the differences in the representation of propositions of varying logical grammatical complexity disappear. When a lesion involving particular language abilities affects the brain, as in the case of patients with left posterior lesions, these differences may reappear.

The results also suggest that in general, the amount of information contained in propositions affects translation to an internal representation more than their logical grammatical complexity does. This explanation is confirmed by the analysis of question recognition scores, where error rates for relation questions were remarkably higher than for assignment questions.

As to our second hypothesis, a severe impairment associated with frontal lobe damage was found in the recognition and matching of propositions and questions. When we take into account the good predictive value of almost every type of proposition and question for arithmetical problem solving ability, these results also suggest that in comparison with healthy controls, the impaired performances of frontal patients in arithmetical word problem solving can be partially attributed to their defective translation skills. In other words, the early stage of translation of text-data into an internal representation already has a significant (negative) effect on the processing of arithmetic word problems. The "defective preliminary orientation" (Luria, 1966) and the "impulsivity" (Christensen, 1975) of frontal patients in arithmetical word problem solving has its origin in a defective encoding of problem sentences and questions, especially of the relational or comparative type.

Our investigation casts some doubts on Barbizet's (1970) interpretation of impaired arithmetical word problem solving after frontal lobe damage as a defective recall of procedures.

The impaired encoding of individual sentences of word problems implies that the frontal patient, in contrast with Barbizet's view, does not possess the information necessary to solve the problem. It is hard to imagine how, with such flawed information at hand, the frontal patient should be able to proceed to a correct solution. Moreover, some descriptive evidence (Luria & Tsvetkova, 1967; Lhermitte et al., 1972) and our observation of the frontal patients behavior during the word problem solving experiment strongly suggests that with their incomplete and erroneous data, frontal patients are quite able to proceed, but that they arrive at what often seem odd answers .

Only one relevant difference was found between the three frontal subgroups: in the recognition task, patients with right frontal lesions performed significantly better than patients with bilateral lesions. This indicates that our hypothesis concerning the better verbal memory performance of the right frontal group in comparison with the left and bilateral frontal groups is not completely confirmed. Although right frontal patients perform better than patients with bilateral and left frontal lesions, the difference with the left frontal group does not reach significance. Apparently, the greater mass effect of bilateral frontal damage produces a more severe impairment in verbal memory functioning than does unilateral (left) damage.

In the recognition task, the number of translation errors made by patients with left posterior lesions is significantly higher than the number of errors made by frontal patients. Such a substantial difference is not found in the sentence-picture matching task. This suggests that for patients with left posterior damage, a representation of word problem sentences at the perceptual level may be better preserved than a representation at the verbal memory level, although both ways of encoding word problem sentences do not reach normal standards.

The higher error rates expected for patients with left posterior lesions in the translation of complex relation and compare sentences are present in the recognition task, but the absence of interaction between groups and complexity indicates that this effect is also present in other groups. This absence of relative increase of error rates in the left posterior group could be due to the nature of our recognition task, in which sentence processing at a rather shallow level may be sufficient. A task requiring deeper processing, e.g. the transformation of sentences in equations (Larkin et al., 1980), would possibly have evidenced this effect better.

Finally, our results concerning the relation between sentence processing and word problem solving suggest that high error rates in almost every type of proposition and question are significant predictors for low problem solving skills, which confirms our final hypothesis, namely that sentence representation should be a major stage in arithmetical word problem solving.

The results obtained in our three tasks provide evidence that frontal and left posterior damage to the brain affects the stage of sentence encoding in arithmetical word problem solving. They do not support the notion that the logical grammatical sentence complexity influences the encoding difficulty, unless there is a brain lesion affecting language abilities. The length of a sentence and the amount of information it conveys seem to determine the encoding difficulty more decisively.

What these results obviously cannot reveal are impairments in later stages of word problem solving. These stages could be a future issue of study, in order to specify the broad theoretical concepts introduced by Luria (1966) and Christensen (1975) further. Such research would also open the possibility to use remedial interventions targeted at particular stages in the information processing chain, as already proposed by De Corte & Verschaffel (1987) and Mayer (1987) for children with arithmetical word problem solving difficulties.

CHAPTER V

The Categorization of Arithmetic Word Problems by Normals, Frontal and Posterior-Injured Patients*

5.1. Introduction

The frontal lobes play a basic role in many forms of human behavior, especially with regard to the regulation of complex activities (Luria, 1966; Walsh, 1978; Lezak, 1982; Stuss & Benson, 1986). A representative example of these complex activities can be found in the resolution of arithmetic word problems. According to Luria & Tsvetkova (1967), arithmetic word problems are "...the most precise and completest model of the intellectual act." In other words, solving arithmetic word problems presupposes the availability of every skill needed to accomplish complex cognitive tasks. Patients with frontal lobe pathology have severe difficulties in solving arithmetic word problems (Luria, 1966; Luria & Tsvetkova, 1967; Lhermitte et al., 1972; Walsh, 1978).

Different theories have been put forward to explain why patients with frontal lobe damage are unable to cope with complex problem solving tasks in

* This chapter is a slightly modified version of L. Fasotti, P.A.T.M. Eling, J. van Houtem (1992). The categorization of arithmetic word problems by normals, frontal and posterior-injured patients. Submitted for publication in Journal of Clinical and Experimental Neuropsychology.

general and with arithmetic word problems in particular. According to Shallice (1982), a cognitive system – the Supervisory Attentional System – located in the frontal lobe, contains the general programming and planning systems that can operate on novel tasks in every (problem solving) domain. Damage to the frontal lobes would leave the patient controlled by contention scheduling, a cognitive mechanism that is only capable of steering cognitive patterns needed for routine behavior. The production of novel non-routine behavior, currently needed in problem solving tasks, would therefore be strongly impaired in the patient with frontal lobe damage. Lezak (1982) also explains the problems of patients with frontal lesions within a general framework. In her view, patients with frontal lesions would have impaired "executive functions" or would lack the cognitive skills necessary for formulating goals, planning how to achieve these goals, and carrying out the plans effectively.

In a detailed description of arithmetical word problem solving after frontal and posterior brain damage, Luria & Tsvetkova (1967) conclude that the primary defect of patients with frontal lesions is their lack of preliminary orientation into the conditions of the problem. Without any preliminary analysis of the different conditions posed by the problem, patients with frontal lobe pathology would just start to perform arithmetical operations with the first fragment that catches their attention. This orientation in the problem-conditions is present in patients with posterior left hemisphere lesions. Nevertheless these patients also fail to solve word problems; they are hindered by their poor understanding of complex sentences and are unable to transform complex logical grammatical constructions into arithmetical operations.

The aim of the present study is not to approach the issue of arithmetical word problem solving from a generally formulated theory of the functional role of the frontal lobes, but to start from a task analysis of arithmetic word problems. This change of focus is consistent with recent work in the field of problem solving, in which the emphasis is on the importance of domain-specific knowledge rather than on the use of general solving strategies (Chi, 1978; Greeno, 1980b; Glaser, 1984; Mayer, 1987).

Moreover, during the past decade, substantial progress has been made in explaining the (domain-specific) human information processes that underlie mathematical word problem solving (Chi et al., 1981; Riley et al., 1983; Briars & Larkin, 1984; Berger & Wilde, 1987). An important issue on which most of these theories agree is that the first main stage in the solution of mathematics word problems consists in the formation of a global internal representation of the problem (Larkin et al., 1980; Mayer et al., 1984; Mayer, 1987; De Corte & Verschaffel, 1987). In this representation stage, the problem solver needs access to different kinds of knowledge in order to encode the problem correctly (Riley et al., 1983; Mayer et al., 1984). Several studies have evidenced that problem-schema knowledge, or knowledge of problem types, is needed to connect the different sentences of a word problem into a comprehensible whole (Mayer et al., 1984; Berger & Wilde, 1987).

An important finding of research on expert and novice performance in mathematical word problem solving is that the recognition of a familiar problem pattern or problem-schema is helpful in providing access to different procedures that are essential in later stages of the solution process. Individuals expert with physics word problems, for example, often have a schema of the structure of the entire problem which allows them to work the problems from the bottom up, taking a direct path toward a solution (Larkin et al., 1980). Expert diagnosticians in medicine represent particular cases by illness-scripts, schemata describing the course and symptoms of a disease, and these scripts facilitate the formation of hypotheses during the diagnostic process (Schmidt et al., 1990). Children, lacking the appropriate schemata of certain types of arithmetics word problems, have difficulties in solving these problems (Greeno, 1980a; Riley et al., 1983). Moreover, schema acquisition facilitates solving in different categories of problems (Cooper & Sweller, 1987).

Therefore, in the present study a sorting task will be used in order to determine the kinds of categories three groups (healthy controls, frontal and left posterior-injured patients) impose on a series of arithmetic word problems. A quantitative as well as a qualitative analysis will be performed on the classifications produced by each group.

5.2. Subjects

Subjects were 16 healthy controls, 30 patients with frontal lobe damage and 12 patients with left posterior pathology.

For every patient the localisation of the lesion site was determined following a procedure introduced by Mazzocchi and Vignolo (1978). This procedure, which consists in the mapping of of a brain lesion as shown by a CT-scan on a standard lateral diagram of the hemispheres, was used with 35 patients. For 7 other patients, extensive CT-scan reports by a radiologist were available.

The selection was based on the criteria that frontal lesions should be situated anteriorly to the Rolandic fissure, and the extension of posterior lesions should be limited to areas behind this fissure. The lesion site in left posterior patients could include parieto-occipital or parieto-temporal brain tissue.

Patients with damage extending both in front of and behind the central sulcus were left out of the study, just as patients with a severe form of aphasia or with reading difficulties. Aphasia was mostly assessed by the Aachen

Aphasia Test (Huber et al., 1984) and in a few patients with posterior lesions by the Token Test (De Renzi & Vignolo, 1962). Moderate comprehension problems were found in 3 patients with left posterior lesions. As these patients had no problems with the comprehension of task instructions, they were not excluded from the study.

This selection procedure left us with 42 patients, distributed as follows between 4 groups: 10 subjects had bilateral frontal damage, 10 had left frontal lesions, 10 right frontal and 12 left posterior lesions.

The etiology of our brain-damaged patients was traumatic in 12 patients, vascular in 14, and tumoral in 16.

The selection of healthy controls was based on an educational level corresponding with an IQ of approximately 100, following the estimation procedure recommended by Luteijn & van der Ploeg (1983) for the Dutch population.

Relevant patient and control characteristics are given in Table 5-I.

Table 5-I: Patient and control profiles.

		bilateral frontal (n=10)	left frontal (n=10)	right frontal (n=10)	left posteror (n=12)	controls (n=16)
sex	male	7	6	7	7	11
	female	3	4	3	5	5
age	m	37.8	42.2	28.9	47.6	26.8
	s.d.	12.6	9.7	9.2	8.0	6.4
WAIS-R PIQ	m	101	109	107	98	104
	s.d.	16	12	19	12	11
WAIS-R VIQ	m	98	105	110	87	104
	s.d.	15	14	16	22	11

5.3. Procedure

Every subject participating in the experiment was presented a deck of 24 small filing cards. On each card an arithmetic word problem was printed. The cards contained 4 kinds of evenly distributed (6 per sort) arithmetic word problems. The characteristics of the different word problems are illustrated in Table 5-II.

Table 5-II: Examples of source formulas and word problems for 4 different types of arithmetic word problems.

Family	Example of source formula	Example of word problem
1. Amount calculation (word problems nr. 1 to 6)	Total amount = amount 1 + amount 2	A train pulls 11 wagons. In a station another 14 wagons are attached to the train. How many wagons does the train pull when it leaves the station?
2. Rest calculation (word problems nr. 7 to 12)	Rest quantity = total quantity – consumed quantity	John has 12 books on a shelf. He has read 8 of them till now. How many books to read are there left on the shelf?
3. Distance calculation (word problems nr. 13 to 18)	Distance = rate : time	Ann walks 15 minutes from home to work, at a mean speed of 8 km. per hour. What is the distance from her work to home?
4. Cost calculation (word problems nr. 19 to 24)	Total cost = unit cost x number of units	The actual price of milk is 0.44£ per liter. Eric buys 5 liter milk in a shop. How much does he pay?

Every subject was given the following instruction: "Here you have 24 arithmetic word problems. Every word problem is printed on a different card. You don't have to solve the different problems. Please sort the problems into groups, based on whatever criteria that you believe are relevant. Word problems that you feel belong together should be assigned to the same group. You can make as many groups as you think are necessary. I give you an example : "A wheat field is 42 feet long and 56 feet large. What is the total surface of this field ?". This word problem is clearly related to surface calculation and is of another type than the following problem: "A farmer collects 450 eggs per day. He breaks 2% of them. How many eggs does he sell on the market each day ?". In the last problem it is not surface, but percentage calculation that is important. Do you see the difference ? At the end you will have several stacks of cards with different types of problems. Name the classification principle on which you have based each stack."

5.4. Scoring and data analysis

Every subject produced a certain number of stacks with a certain number of items. Three analyses were performed on these data:

1. For each group, the mean number of produced categories was calculated.
2. The frequency of occurence of the labels given to each stack was computed per group.
3. For every group, the frequency of occurence of 2 items in the same stack was calculated. This information allows the construction of a 24x24 similarity matrix. Each cell represents the frequency with which 2 items coincide in the same stack. Depending on the groups, these frequencies can vary between 0 (pairs of items never placed together) and 16 (pairs of items always placed in the same stack by normals), 30 (pairs of items always placed in the same stack by frontal patients) or 12 (pairs of items always placed in the same stack by patients with left posterior lesions). The statistical techniques used to analyse the data were multidimensional scaling (MDS) and hierarchical cluster analysis (HCA).

MDS is a procedure aimed at the discovery of the latent structure of a data pool, when the degree of similarity between the items is known. MDS differentiates the items in various dimensions. The result is a multidimensional representation of the items in the form of a geometrical configuration of points. In this configuration, items with great similarity are put closer to each other. Larger distances correspond to less similarity (Kruskal & Wish, 1978). In the MDS-procedure used in the present study (ALSCAL), the number of dimensions was chosen on the basis of Young's formula, which minimizes s-stress.

HCA is aimed at grouping items with a relatively high degree of similarity. Each frequency matrix was submitted to HCA, using the single linkage method. The number of clusters for each group can be determined by inspection of the fusion coefficients. The agglomeration of items was stopped as soon as the gap between two adjacent steps became large in comparison with the other increases. For each group, the clusters obtained in HCA were embedded in the geographical MDS representation.

5.5. Results

An examination of the control groups' data reveals very consistent word problem sorting performances. The mean number of categories produced by this group was 4.1 (s.d.= 1.12) and the categories based on a source formula or a family (see Table 5-II) captured 70% of the word problems. An analysis of the labels given to the different stacks shows that the following labels were used by 50% or more of the 16 subjects in the control group:

Addition	(13 x)
Subtraction	(12 x)
Multiplication	(8 x)
Distance calculation	(8 x)
Division	(7 x)

Another 13 different labels were used by one or two subjects at most, indicating that categories who were not based on an equation or a family were uncommon in this group.

The dimensions on the basis of which controls sorted the 24 word problems were analysed by means of the statistical procedure ALSCAL (Fig. 5-1). Using Young's s-stress formula, 2 dimensions proved to be most appropriate for the interpretation of the results. We felt we could assign the label "operation" to dimension 1. In this dimension, complex multiplication and division word problems are found at one pole, whereas less complex addition and subtraction problems are present at the other pole. The second dimension was labelled "initial quantity". Examination of this dimension reveals that word problems in which an initial quantity is increased are found on one extreme, whereas word problems in which an initial quantity decreases are found on the other extreme.

HCA resulted in 4 different clusters. In Figure 5-1 these clusters are embedded in the geographical MDS representation. Based on inspection of the labels given by normal subjects to the stacks, the clusters were labelled as:

1. Addition (items nr. 1-6) 2. Subtraction (items nr. 7-12) 3. Division-distance calculation (items nr. 13-18) 4. Multiplication (items nr. 19-24). In other words, normal subjects were quite able to recognize the structure of the word problems presented and did so with high agreement.

The mean number of categories for the frontal group was 6.0 (sd = 3.12) and categories based on a formula or a family captured only 29% of the sorted word problems.

Two dimensions appeared also sufficient to describe the frontal groups' sorting results. Figure 5-2 shows these dimensions. Dimension 1 (horizontal plane) groups word problems with identical arithmetical configurations. On one end there are word problems that concern the calculation of concrete quantities (marbles, apples, litres of milk). This extreme contains addition as well as subtraction problems. On the opposite end the word problems belong to a more abstract arithmetical configuration or problem family, namely that of distance calculations (distances on highways, walking distances, length of

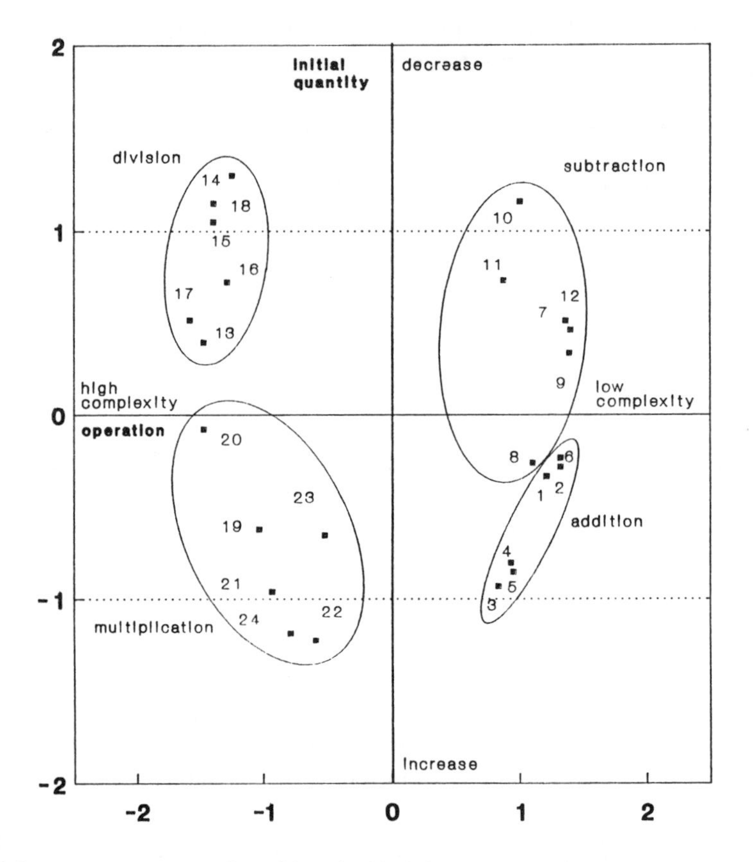

Fig. 5-1: **MDS-representation with embedded clusters of the control groups' sorting performance.**

a wall that has to be painted). The second dimension (vertical plane) is characterized by the objects to which the different problems refer. On one pole the word problems concern concrete objects such as books and trains, whereas the other pole groups more abstract word problems related to financial matters (salaries, money, toll on highway).

The total number of labels used by frontal subjects to characterize the 24 word problems amounted to 89. None of these labels was used by more than 50% of the 30 frontal subjects. The most frequently used labels were:

Rest	(10 x)
Quantity	(8 x)
Money	(9 x)
Distance	(9 x)
Books	(7 x)
Trains	(5 x)

The HCA procedure yielded 10 clusters. One of these clusters contained only one item. In Figure 5-2 the remaining 9 clusters are embedded in the MDS representation. Each cluster was named after the most frequently used label by frontal subjects. Two remarks must be made about this labelling. Firstly, the "rest" label was never used to indicate items belonging to the family of rest calculation problems (see Table 5-II), but was always employed to label a disparate group of items that could not be classified further by the frontal subjects at the end of the sorting task. Secondly, the "money" label was

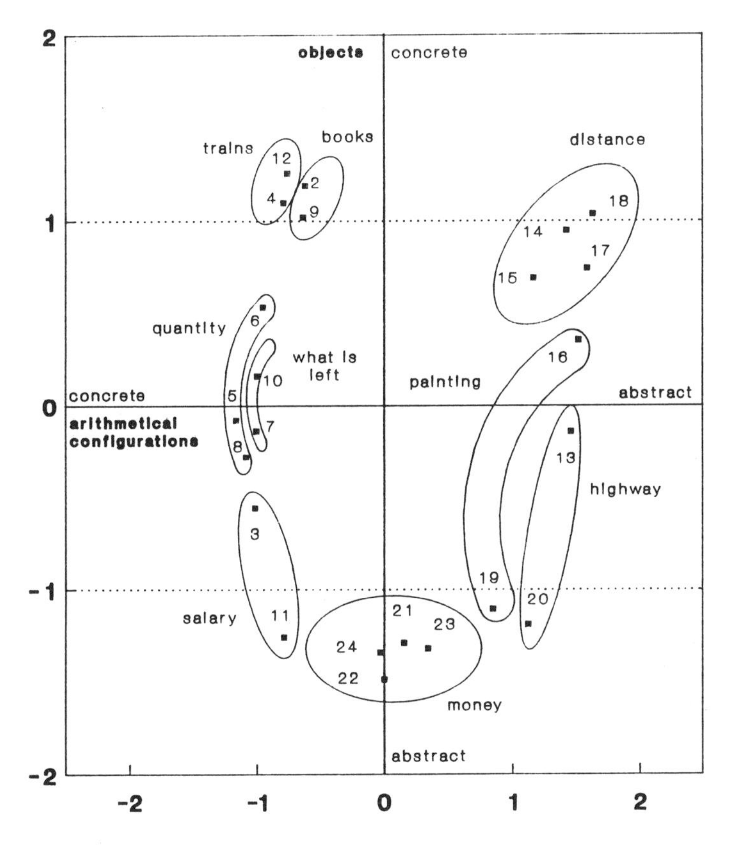

Fig. 5-2: MDS-representation with embedded clusters of the frontal groups' sorting performance.

employed to label cost calculation problems, although word problems involving money were also present in the quantity and rest calculation families. The dimensions as well as the components found in the analysis of the frontal patients' sorting performances appear to be entirely different from those found in the control group. Frontal patients seem to rely on superficial features such as concrete objects whereas healthy controls use such relevant information as an arithmetical operation to sort word problems. Only one cluster (distance) seems to be based on a template that presupposes a deeper processing of the word problem text.

The mean number of categories produced by patients with left posterior lesions was 9.7 (s.d.= 3.6), whereas categories based on equations or families included not more than 10% of the word problems. The interpretation of the left posterior groups' sorting dimensions was difficult. Young's s-stress formula suggested that 2 dimensions could represent the data, but due to the heterogenity of the items grouped on the extremes of these dimensions (see Figure 5-3) an unambiguous labelling of the dimensions was impossible.

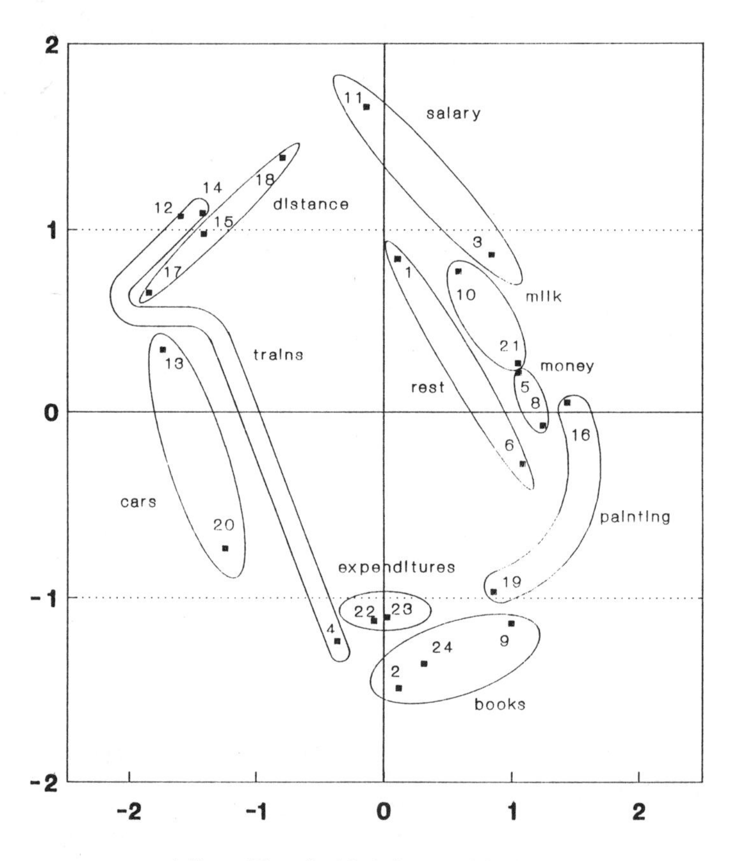

Fig. 5.3. **MDS-representation with embedded cluster of the left posterior groups' sorting performance.**

The total number of labels used by this group was 61. The most frequently used labels were:

Books	(8 x)
Trains	(7 x)
Distance	(7 x)
Milk	(5 x)
Money	(4 x)
Cars	(4 x)
Salary	(4 x)

The HCA procedure produced 11 clusters, one cluster containing only one item. Three clusters were in every aspect identical to those made by the frontal group (salary; highway, cars; painting). Three clusters differed only by a single item (distance; trains; books). Four other concrete clusters (milk; rest; money; expenditures) were exclusively used by the left posterior group. One of these clusters, namely "expenditures', showed strong similarities to one of the clusters of frontal patients. It contained two items that were found in the "money' cluster of the frontal group.

5.6. Discussion

The objective of this study was to investigate quantitative and qualitative differences between healthy subjects, frontal and left posterior brain-damaged patients in an arithmetical word problem sorting task.

A quantitative analysis shows that the mean number of the categories used differs between the groups. While healthy controls create a relatively small number of categories, the number of categories produced by frontal and left posterior patients is significantly higher (Mann-Whitney U test, $z=-1.79, p<.05$ and $z=-3.78, p<.01$ respectively) The left posterior groups' mean number of categories is also significantly larger than the frontal groups' (Mann-Whitney U test, $z=-2.13, p<.01$). In the control group, the categories bases on an equation or a family capture the majority (70%) of the sorted word problems, whereas only a fraction of these problems is captured in the categories created by frontal (29%) and left posterior (10%) subjects.

These data suggest that normals recognize and classify arithmetic word problems according to features that are relevant for problem solving strategies. Frontal and left posterior patients' sorting behavior, on the other hand, is clearly led by a larger and qualitatively different number of sorting principles. The critical question then becomes: what are the bases on which healthy controls and patients categorize word problems?

According to the MDS and HCA analyses, it appears that healthy controls classify the word problems according to the major arithmetical operation governing the solution of each problem. The two major dimensions according to which the problems are sorted represent the complexity of their arithmetical operations and the capacity of these operations to increase or decrease an initial quantity, whereas the clusters simply group together the problems characterized by the same operations. The labelling of the stacks by the healthy subjects confirms this analysis. The most frequent labels refer to arithmetical operations. Only in the case of divisions the family of problems is named more frequently than the operation. If we define the "deep structure" of a word problem as the underlying arithmetical operation applicable to this problem, analogous to Chi et al. (1981), then this deep structure is the basis by which normals group word problems.

Examination of the frontal and left posterior groups' dimensions and clusters reveals that the "surface structure" of the word problems is the basis for classification. Following Chi et al. (1981) "surface structures" can be defined as (a) the objects referred to in a problem (e.g. books, trains, cars, milk), (b) the arithmetical terms referred to in a problem (e.g. what is left) and (c) the arithmetical configuration described in a problem (e.g. expenditures, quantities).

There is a considerable overlap between the clusters created by frontal and left posterior patients. The main difference between both brain damaged groups resides in the dimensions along which the different word problems are sorted. While frontal patients group the problems along the two dimensions of objects and arithmetical configurations, both going from concrete to more abstract items, such clearcut dimensions are not found in the left posterior patients' sorting behavior. Nevertheless, the similarities between both groups are remarkable. Although the frontal group is better at recognizing families of problems (e.g. distances), both groups resemble each other in the use of "surface structures" as a basis for categorization. Surprising is the fact that the left posterior group seems to use more objects than the frontal group as a means for classification. This is in sharp contrast with the ideas of Goldstein (Goldstein, 1959; Goldstein & Scheerer, 1941), that impairments of abstraction are maximal with lesions of the frontal lobe. This contrast may be understandable, when one takes into account the nature of the language impairment of the left posterior group, consisting in the inability to transform logical grammatical constructions into arithmetical operations. Such an impairment could have induced these patients to use simpler and more concrete language categories for classification. Moreover, there is some evidence that patients with posterior lesions – especially on the left – frequently perform as poorly as frontal patients in sorting or categorizing (Teuber, 1964).

For frontal and left posterior brain-damaged patients the lack of recognition of a pattern or a schema for types of problems may have two consequences for word problem solving. It directly influences the interpretation of word problems, by weakening the distinction between important and unimportant information for problem solution. The results of our experiment suggest that frontal and left posterior brain-damaged patients wrongly judge superficial features of a word problem as being very essential. Secondly, just as in novice problem solving performance, poor problem-schema understanding can impede quick access to relevant procedures for the solution of word problems. This would oblige the patient, just as the novice solver, to work backwards from the problem goal to the given quantities, whereas expert solvers usually work forward from the givens to the desired quantity.

In this study the arithmetical word problem sorting behavior of normals, frontal and left posterior-injured patients was investigated. In order to obtain a balanced group of frontal subjects, patients with right as well as with left and bilateral lesions were included in the frontal group. Due to the smallness of the samples, the statistical techniques used in the present study could not be applied to each frontal subgroup separately. However, in future research it may be interesting to use larger groups in order to investigate if there are significant differences in word problem sorting behavior between frontal patients with different hemispheric lesion location.

Language impairments were controlled for in each brain-damaged group. No explicit aphasic symptoms could be found in the frontal group, whereas subjects with left posterior lesions showed only mild comprehension problems which did not prevent them to understand task instructions, as far as could be assessed by clinical observation. The current results do not allow to conclude whether the sorting behavior of brain-damaged subjects can be attributed to a sorting problem or whether an eventual language impairment of a higher order is also involved. In future research, a more comprehensive use of language comprehension tasks should permit to decide on this issue.

CHAPTER VI

Influence of Improved Text Encoding on Arithmetical Word Problem Solving after Frontal Lobe Damage*

6.1. Introduction

The purpose of this study is to explore some specific aspects of arithmetical word problem solving that could be improved in patients with frontal lobe lesions. Following recent developments in the field of mathematical problem solving (Mayer, 1983; Briars & Larkin, 1984; De Corte & Verschaffel, 1987; Mayer, 1987) we will examine whether a cueing procedure aimed at improving two basic encoding skills, sentence translation and problem-schema understanding, can enhance the arithmetical word problem solving performance of patients with frontal and left posterior brain damage.

According to several theories of mathematical problem solving (Larkin et al., 1980; Mayer et al., 1984; De Corte & Verschaffel, 1987) the first stage in word problem solving is problem representation. At this stage, the problem solver tries to properly encode the text of a word problem. This encoding process consists of (a) a translation of each sentence of the problem into an internal representation and of (b) the integration of the information from the

* This chapter is a slightly modified version of L. Fasotti, J.J.C.B. Bremer, P.A.T.M. Eling (1992). Influence of improved text encoding on arithmetical word problem solving after frontal lobe damage. Accepted for publication in Neuropsychological Rehabilitation.

different sentences into a coherent representation of the problem (Mayer et al., 1984; Mayer, 1987).

Although more globally conceptualized, the poor encoding of problem givens has been anecdotically reported as a basic factor contributing to the poor performances of patients with frontal lobe damage in arithmetical word problem solving. Luria & Tsvetkova (1967) have stated that much of the difficulties of frontal patients arise through an incomplete analysis of the givens of arithmetic word problems and the tendency of these patients to perform arithmetical operations only on the first fragment of the problem that catches their attention. A similar impulsive behavior is clearly described in a single case study by Lhermitte et al. (1972). The patient in question had an occlusion of the anterior cerebral artery causing damage to the left frontal hemisphere. His arithmetical problem solving behavior is described as follows: "...After reading the problem he immediately writes down the three numbers appearing in the givens (of which, by the way, one is wrong), after which he adds the numbers without even taking into account the minus sign attributed to the last number." (p. 431). More descriptive evidence comes from Christensen (1975), who concludes that patients with frontal lobe damage only grasp isolated fragments of arithmetic word problems on which to operate, without reckoning that these fragments cannot lead to a subgoal or to the final goal of the problem.

Patients with lesions of the left posterior hemisphere (in parieto-occipital or parieto-temporal areas) also experience marked difficulties in the encoding stage. The cause of their problems, however, is assumed not to lie in impulsive behavior, but in their incapacity to understand sentences containing words expressing relations (e.g. ... sells 12 *less* than), complex prepositions (e.g. ... sells 23 *per* day) and conjunctions (e.g. ... *if* he sells...). This disturbance of comprehension has occasionally been referred to as "semantic aphasia" (Luria, 1966; Christensen, 1975; Kertesz, 1979).

These neuropsychological descriptions have focused on the different types of encoding impairments during arithmetical word problem solving, whereas research in mathematical problem solving has directed its attention to the mechanisms that underly correct encoding.

The results of these investigations indicate that within the encoding stage of word problem solving, the comprehension of sentences (and their simultaneous translation into a semantic representation) is a condition for correct word problem solving (Soloway et al., 1982; Riley et al., 1983; Kintsch & Greeno, 1985; Bovenmayer Lewis & Mayer, 1987). The same can be said of schema understanding, i.e. the knowledge of problem types, that allows the solver to categorize a problem properly. It appears that, when subjects lack an appropriate schema for mathematical word problems, problem representation is more likely to be faulty (Larkin et al., 1980; Mayer, 1982; Berger & Wilde, 1987).

If both these encoding skills are a prerequisite for correct word problem solving, then it seems worthwhile to investigate if sentence translation and schema understanding can be improved, and whether this would result in an improved word problem solving ability. An experiment was therefore performed, in which patients with frontal and posterior lesions were asked to solve two kinds of arithmetical word problem tasks. The first type of task comprised a series of 8 arithmetic word problems of increasing complexity, that had to be solved after they had been read aloud by the subject. These problems will be referred to as "uncued problems". The second set of arithmetical word problem series was analogous to the first, but after reading each problem, the subject had to answer a number of questions dealing with sentence comprehension and schema knowledge. In order to improve correct encoding, verbal feedback was provided after every question, whereupon the subject was asked to solve the problem. These word problems are referred to as the "cued problems".

This procedure was used in order to verify the following three hypotheses about the word problem solving performances of patients with frontal and left posterior brain damage.

1. In solving uncued arithmetic word problems, the performances of both frontal and left posterior groups should differ significantly from those of a group of healthy controls.
2. In solving cued arithmetic word problems, the performance of patients with frontal lobe lesions should improve significantly. As our cueing procedure was aimed at building a useful internal representation of word problems, it should reduce impulsive behavior and thereby enhance correct encoding. Better encoding should, in its turn, improve word problem solving ability.

   The performance of patients with left posterior lesions, on the other hand, should not differ significantly from uncued to cued word problems series. Our cueing procedure was aimed at reflection upon problem givens and not at the amelioration of a language impairment. Consequently, we expected little improvement from uncued to cued conditions for left posterior patients.
3. If impulsivity indeed leads to a hasty and incomplete reading and processing of text data, the changes in performance in the frontal groups after cueing should affect solving performance as well as solving time. In these groups, the better encoding of text data should not only enhance the overall solving performance, but also improve the availability of data on which to operate, and thus increase the time needed to find a solution. The solving

times of patients with left posterior lesions should exceed the times needed by frontal patients in both cued and uncued solving conditions. This hypothesis was based on Luria and Tsvetkova's supposition that "...the comprehension of a grammatically and logically complex text is impossible for them (note author: left posterior patients) and the work that a normal subject executes in a few seconds takes them an unduly long time without producing any result indeed." (Luria & Tsvetkova, 1967). This description also implies that there should be no significant differences in solving time between uncued and cued arithmetic problem series in the left posterior group.

To assess the transfer of eventual learning effects from the cueing procedure, a second word problem solving session was organized after an interval of 2 to 4 days. In this session, analogous uncued and cued word problem series were presented to every subject.

6.2. Subjects

Fifty subjects participated in this study: 30 patients with frontal lobe damage, 10 patients with left posterior hemisphere lesions and 10 healthy controls. In the frontal group 10 patients had left hemisphere lesions, 10 had right hemisphere lesions and 10 had bilateral frontal lobe damage. The site of the different lesions was defined as follows: frontal, if the damage was confined to cortical-subcortical areas located anterior to the Rolandic fissure; left posterior, if the damage was situated behind the Rolandic fissure in the left hemisphere. In the latter group the lesion had to concern the parietal lobe, either in conjunction with temporal or with occipital areas.

CT-scans of 27 patients with frontal lesions and of 6 patients with left posterior damage were available. For all these patients the site and extent of lesion shown by the scan was mapped on a standard lateral diagram of hemisphere, according to a procedure described by Mazzocchi & Vignolo (1978). For the remaining 7 patients (3 with frontal and 4 with left posterior lesions) data with regard to the site of the lesion were gathered from extensive CT-scan reports drawn up by radiological experts.

The etiology of the different brain lesions was of tumoral origin in 15 cases, of traumatic origin in 12 and of vascular origin in 13 cases.

In the frontal group, patients with a severe aphasia had been excluded on the basis of the results of the Aachen Aphasia Test (Huber et al., 1984). In the posterior group the Tokentest (De Renzi & Vignolo, 1962) revealed that 3 subjects had moderate comprehension problems, which, however, did not hinder the understanding of task instructions.

The selection of healthy subjects for the control group was based mainly on educational level. This level had to correspond with an IQ of approximately 100, in accordance with the estimation procedure recommended by Luteijn & van der Ploeg (1983) for the Dutch population. Care was also taken to select subjects that did not practise regular mathematical activities in their professional or educational settings.

Relevant characteristics regarding age, sex and IQ-levels of experimental groups and healthy controls are given in Table 6-I.

Table 6-I: Patient and control profiles.

		bilateral frontal (n=10)	left frontal (n=10)	right frontal (n=10)	left posterior (n=10)	controls (n=10)
sex	male	7	6	7	5	6
	female	3	4	3	5	4
age	m	37.8	42.2	28.9	46.8	33.3
	s.d.	12.6	9.7	9.2	8.7	6.4
WAIS-R PIQ	m	101	109	107	100	104
	s.d.	16	12	19	11	11
WAIS-R VIQ	m	98	105	110	79	104
	s.d.	15	14	16	15	11

6.3. Procedure

All patients participating in the experiment were requested to solve 4 series of analogous arithmetic word problems. Each series consisted of 8 word problems of increasing complexity. Complexity was defined as the number of arithmetical operations required to solve each problem. The word problems of every series were presented in the same randomized order to each patient.

For each correctly identified step and its corresponding operation one point was given. Errors in the execution of calculations were not counted. The maximum score over 8 word problems in every series was 28 points. This variable is referred to as "performance".

In order to determine normal levels of word problem solving performance, the first series of problems was submitted to our 10 control subjects.

The 4 series submitted to the patient group contained word problems of analogous complexity; every problem had a counterpart in the other series, requiring the same number of operations for solution. The differences between equally complex problems in the various series bore on 4 aspects:

1. the names of the subjects appearing in the problem
2. the objects occurring in the problem
3. the operations involved to solve the problem: addition-subtraction and multiplication-division operations alternated between series
4. the numbers with which to operate varied from 3 to 30.

Table 6-II illustrates an example of 4 equally complex problems from 4 different series.

Table 6-II: Example of equally complex arithmetic word problems in 4 different series.

Peter has 10 marbles. His sister has 21 marbles more. Peter's brother has 5 marbles less than Peter and his sister together. How many marbles do the three children have together?	uncued 1
John is 12 years old. His father is 28 years older. John's mother is 4 years younger than John and his father together. How old are the three together?	cued 1
Stan possesses 26 albums. His son has 11 albums less than him. Stan's friend has 7 albums more than Stan and his son together. How many albums do the three possess together?	uncued 2
Kevin has scored 12 points in a game. His brother has scored 21 points more than him. His sister has scored 9 points less than Kevin and his brother together. What is the total score of the family?	cued 2

In the first series, each problem was presented on a file card and subjects were instructed to read the problem aloud. Thereupon, subjects were requested to solve the problem (still aloud) and to write down all the operations required for solution. Time was recorded starting from the moment that subjects had read the problem until they notified that they were ready. This variable is called "solving time". In table 6-III the problems of the first series are given, together with the number of operations per problem (representing complexity) and the total number of operations for the complete series. This series is referred to as "uncued".

Table 6-III: First series of uncued arithmetic word problems.

	WORD PROBLEMS	NUMBER OF OPERATIONS
1.	Paul has been jogging for 11 miles Frank has been jogging 4 miles less than Paul. How many miles have they jogged together?	2
2.	Ellen has 5 apples. Karen has 4 times more apples. How many apples do they have together?	2
3.	Company A has 20 employees. Company B has 3 employees less than company A. Company C has 5 employees more than company A. How many employees do the 3 companies have in total?	3
4.	Grandmother A has 18 grandchildren Grandmother B has 7 grandchildren less than grandmother A. Grandmother C has one-third of the number of grandmother A's grandchildren. How many grandchildren do the grandmothers have in total?	3
5.	Peter has 10 marbles. His sister has 21 marbles more. Peter's brother has 5 marbles less than Peter and his sister together. How many marbles do the 3 children have together ?	4
6.	Peter is 12 years old. His brother is 5 years younger. His grandfather is 4 times older than Peter and his brother together. How old are the 3 together?	4
7.	An has 22 $. Maud has 17 $ less. Sarah has 9 $ more than An and Maud together. How many $ does every girl have on average?	5
8.	A shopkeeper has sold 11 bottles of milk. A second shopkeeper has sold 3 bottles less. A third shopkeeper has sold 4 bottles more than the first and second shopkeepers together. Together the 3 shopkeepers have 11 bottles of milk left. How many bottles did they buy together?	5
	Total score	28

In the second series, after reading each problem, the patients were provided with cues about the problem sentences and the problem-schema, following a training procedure introduced by Mayer (1987). The procedure consists of confronting the patients with 3 kinds of questions before allowing them to solve the problem. Two sets of questions concern problem sentences; one set is about problem propositions, and a separate set concerns a question proposition. Both sets are used in order to cue sentence translation. The third type of question regards the schema underlying the problem. After every question, verbal feedback about the correctness of the subjects' response was given. Only one answer per question was accepted and no other supplementary information was provided. Table 6-IV illustrates the procedure with one of the word problems of this series, referred to as the "cued series".

Table 6-IV: Cueing of sentence translation and problem understanding in an arithmetic word problem.

---

1. PROBLEM PRESENTATION
   Instruction: "Would you please read this word problem aloud?"

   John is 12 years old.
   His father is 28 years older.
   John's mother is 4 years younger than John and his father together.
   How old are the three of them together?

---

2. CUEING PROCEDURE
   Instruction: "Before solving the word problem answer the following questions."

1. *Which of the following sentences is true?*
   (a) John and his father together are 4 years older than John's mother.
   (b) John's father is 28 years older.
   (c) John and his father differ 28 years.
   (d) John's mother has the same age as John and his father together.

2. *What are you being asked to find?*
   (a) The total age of the family?
   (b) The age of John?
   (c) The difference between the mother's age and John's and his father's age together?
   (d) The difference between John's and his father's age?

3. *One of the following word problems is of the same type as the problem you have to solve. Which one is it?*
   (a) Peter, his father and his sister are 83 years together.
   Peter's father is 41 years old. How old are Peter and his sister together?
   (b) David has 16 marbles. His brother has 19 marbles more.
   His sister has 8 marbles less than David and his brother together.
   How many marbles do they have together?
   (c) Donna buys 16 sweets. Ann buys 6 sweets more then Donna.
   Susan buys 4 sweets less than Ann. The three girls eat 11 sweets.
   How many sweets are left?
   (d) Three farmers have bought 85 kg. of seed. The first farmer uses 16 kg.
   The second 29 kg. How many kg. of seed are left for the third farmer?

---

3. SOLVING
   Instruction: "Can you solve the problem please?"

After the cueing procedure, the patients were asked to solve the word problems exactly as in the first series.

Series 1 (uncued) and 2 (cued) were always carried out consecutively in a single session. The series 3 (uncued) and 4 (cued) were presented to the subjects in a second session, after an interval of at least 2 and at most 4 days after the first.

Although the experiment's design may resemble an ABAB design with one treatment procedure, the second session (series 3 and 4) was mainly introduced in order to see whether eventual gains of the first session (from series 1 to series 2) were lost, maintained or even consolidated after a short inter-period time-interval.

6.4. Results

Before investigating the hypotheses of the study, and in order to control for processing load, an analysis of the patients' performances in function of the level of complexity of the proposed problems was carried out. Therefore, the proportion of correctly solved uncued word problems for every level of complexity was cast into a two-way table having 40 rows (patients) and 4 colums (levels of complexity). The data in each row were ranked from 1 to 4. With the use of these values, a Page test for ordered alternatives was carried out. This test showed that the patients' performance is inversely related to the level of complexity of the word problems (α=.01), confirming the assumption that the number of operations in the used word problems is a significant measure of complexity.

The problem solving performance of our patient groups in the four series of arithmetic word problems is illustrated in Figure 6-1.

The results of a three-way ANOVA for repeated measures with groups, sessions and cueing as factors are given in Table 6-V.

Neither the main effect of sessions, nor its interaction with other variables was significant. It may therefore be safely assumed that for each patient group, baseline and treatment measures were not significantly different in successive sessions. For this reason, baseline and treatment measures were averaged over sessions before testing hypothesis 1.

Figure 6.2 shows the averaged baseline and treatment scores for different patient groups. To illustrate normal levels of performance, the baseline score of the healthy control group is also included in the diagram.

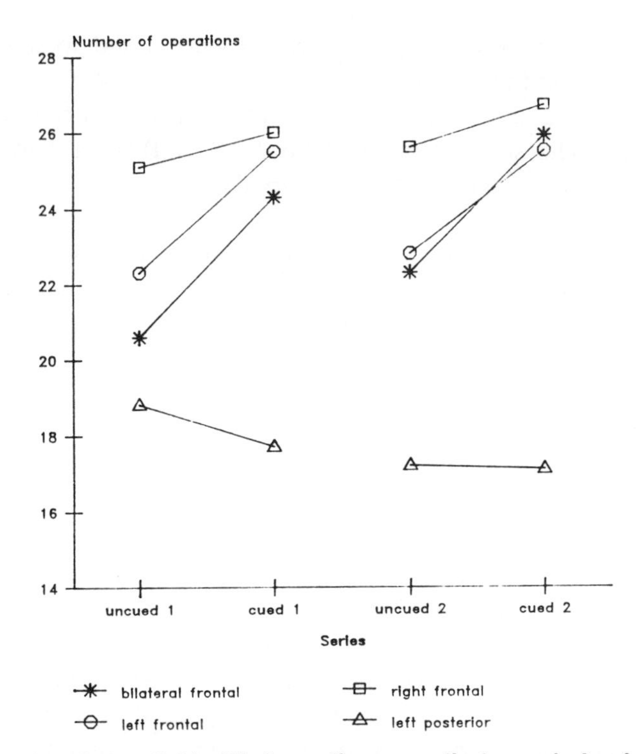

Fig. 6-1: Number of correctly identified operations per patient group in 4 series of arithmetic word problems.

Table 6-V: Summary of analysis of variance.

SOURCE	df	ss	ms	f
Between subjects				
Groups (A)	3	1480.33	493.44	4.81*
Error between	36	3639.95	102.69	
Within subjects				
Sessions (B)	1	4.90	4.90	.90
A x B	3	38.65	12.88	2.37
Error within	36	195.95	5.44	
Within subjects				
Cueing (C)	1	122.50	122.50	21.34*
A x C	3	111.35	37.12	6.47*
Error within	36	206.65	5.74	
Within subjects				
B x C	1	.23	.23	.09
A x B x C	3	3.02	1.01	.39
Error within	36	34.25	2.62	

* $p < .01$

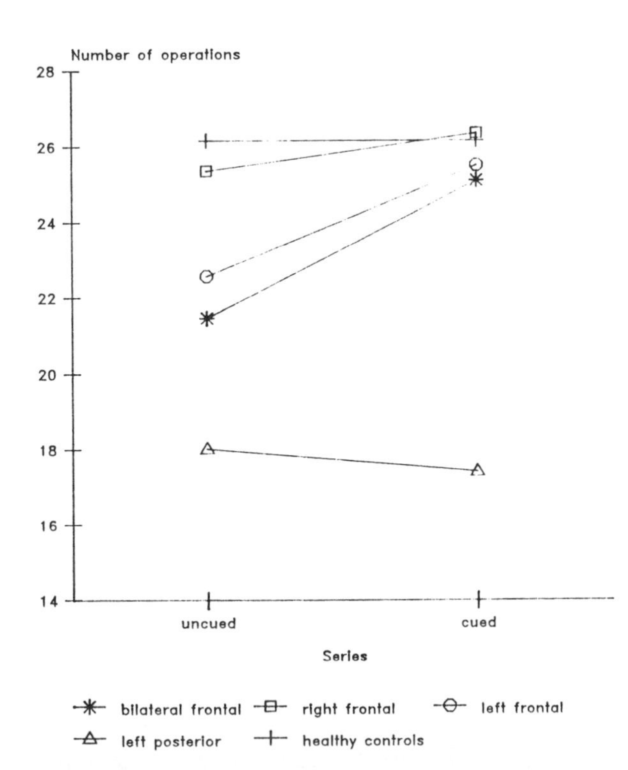

Fig. 6-2: Averaged baseline and post-treatment performances per patient group.

Hypothesis 1 concerned the different performances of brain damaged groups and healthy controls in uncued problem solving conditions. To test this hypothesis, a one-way ANOVA was performed on the baseline scores of Figure 6.2. This ANOVA shows a significant groups effect (F=4.39, df=4/45, p<.01), suggesting different levels of performance in our five groups. Contrast analyses reveal that patients with bilateral frontal (T=-3.35, df= 10.5, p<.01) and left posterior lesions (T=-3.21, df= 9.4, p<.01) perform significantly below normal levels in uncued word problem solving. The difference between the healthy control group and our patients with left frontal lesions tends toward significance (T=-1.98, df= 9.9, p=.07). There is no significant difference between right frontal patients and normals (T=-.86, df= 12.8, p=.41).

Hypothesis 2, related to the changes in problem solving performance after cueing, was examined as follows. Table 6-V shows a significant main effect of cueing and a groups x cueing interaction. These effects indicate that cueing influenced our patient groups differently. A further analysis revealed that, as predicted, the effects of cueing were significant for bilateral frontal (F=23.21, df= 36/1, P<.01) and for left frontal (F=15.16, df= 36/1, P<.01) groups, whereas the performance of the left posterior group remained unchanged (p=.15). The performance of right frontal patients did not improve significantly after cueing (p=.20), but was already at normal level in uncued conditions.

To assess the performance in the patient groups after cueing in comparison with normal performance, a one-way ANOVA was performed on the treatment scores of Figure 6-2. A significant groups effect was found (F=7.23, df= 4/45, P<.01), suggesting that after cueing there are still differences between groups. Multiple contrasts reveal that the three frontal groups perform at normal levels (T=-1.0, df= 11.9, p=.34 for bilateral frontal, T=-.49, df=10.8, p=.63 for left frontal and T=.33, df= 17.7, p=.74 for right frontal patients respectively). Patients with left posterior lesions continue to perform poorly in comparison to normals (T= -3.27, df= 9.4, p<.01). These effects are in line with the patients' spontaneous verbal behavior during the execution of our word problem solving tasks. Patients with left posterior lesions frequently complained that the cueing procedure only increased the confusion and the sense of misunderstanding they felt, whereas such remarks were never heard from subjects with frontal lobe lesions.

Hypothesis 3 stated that the solving times of the frontal groups, who performed better in cued conditions, should increase significantly, that left posterior patients should be significantly slower than frontal patients in all conditions, and that there should be no significant difference in solving time between cued and uncued problem series in the left posterior group. The mean solving times of our 4 patient groups in uncued and cued conditions in the 2 different sessions are given in Figure 6.3. In order to investigate the first two of the above-mentioned hypotheses, a three-way ANOVA with groups, sessions and cueing as factors was conducted on the mean solving times of Figure 6.3, with tests of simple effects within the different patient groups. In none of these groups the effect of sessions was significant (p>.10 in each group), whereas the effect of cueing on solving time was only significant in the bilateral frontal group (F=16.11, df=1/29, p<.01). The left frontal group, the other group performing significantly better in cued conditions, also shows longer mean solving times after cueing, but the difference is slight in both sessions and does not reach significance (p=.17). The interaction between sessions and cueing can be accounted for by chance alone in each frontal group (p.>05), indicating that for every frontal group changes in solving time after cueing do not differ significantly in both sessions.

To test Luria & Tsvetkova's idea that patients with left posterior lesions should be slower than frontal patients in every condition, two separate one-way ANOVAS were performed on the mean solving times of the cued and the

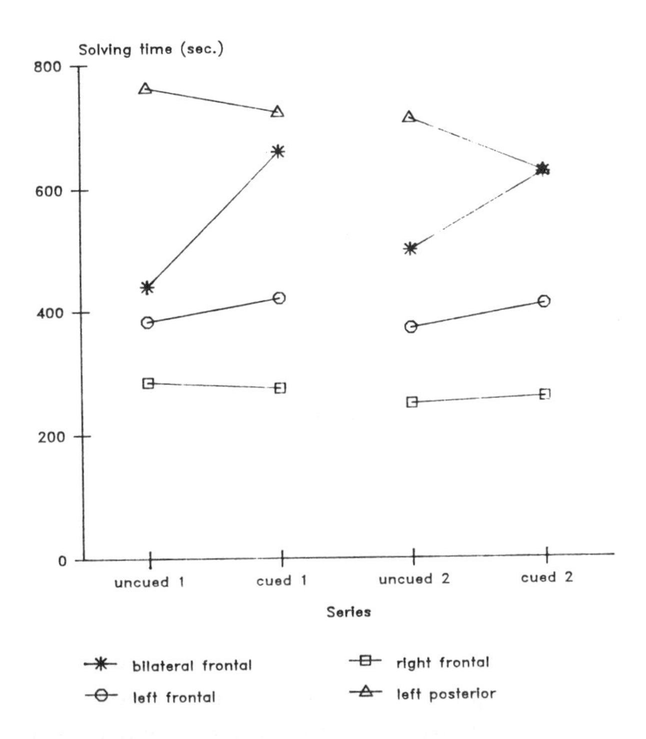

Fig. 6-3: Solving times of patient groups in 4 series of arithmetic word problems.

uncued word problem series. Significant group effects were found in the uncued (F=6.68, df=3/30, p<.01) as well as in the cued (F=5.57, df= 3/31, p<.01) condition. Further analyses with multiple contrasts revealed that the solving times of the left posterior group in the uncued conditions are significantly higher than those of right frontal (T=-3.79, df= 9.9, p<.01) and left frontal (T=-2.74, df= 9.9, p<.05) patients. The difference with bilateral frontal patients tends toward significance (T=-.19, df= 13.0, p=.07). After cueing, the increased solving times of bilateral frontal patients reduce this difference significantly (p=.80), whereas the left posterior group remains significantly slower than the right (T=-4.41, df= 13.5, p<.01) and left frontal (T=-2.16, df=15.8, p<.05) groups. These results indicate that patients with left posterior lesions spend considerably more time than frontal patients trying to understand word problem sentences. Only in cued conditions, the bilateral frontal group is as slow as the left posterior group. Both groups remain nevertheless different: whereas in cued conditions patients with bilateral frontal lesions improve their solving performance at the expense of solving time, patients with left posterior lesions remain slow and perform poorly at the same time. Finally, the effect of cueing on solving time is not significant in the left posterior group (p=.44), confirming the last part of hypothesis 3, which stated that left posterior patients should be equally slow in uncued and in cued problem solving conditions.

6.5. Discussion

The aim of the present experiment was to investigate the differences between frontal patients, patients with left posterior lesions and healthy controls in arithmetical word problem solving. A cueing procedure, aimed at improving encoding by inhibiting impulsiveness, was used to ameliorate the solving performance of frontal patients. Besides word problem solving performance, solving time was also studied as a dependent measure.

Our first hypothesis stated that, in uncued solving conditions, healthy controls should perform better than brain damaged patients. Our findings indicate quite clearly that word problem solving is impaired in every brain damaged group, except for patients with right frontal lesions. This group of patients was able to reach normal levels of performance.

Two explanations might be considered for this finding. On the one hand, linguistic abilities after right frontal damage could be more spared than after left or bilateral frontal lesions. This view is advocated by several authors (Benton, 1968; Kaczmarek, 1984; Novoa & Ardila, 1987). As a consequence, the ability to handle the complex logical grammatical constructions in word problems could be better preserved in patients with right frontal lesions. On the other hand, a better representation of word problem sentences could facilitate the planning of subgoals and the identification of the operations that are necessary to solve these subgoals. This claim is supported by evidence that patients with right frontal lesions can perform quite normally in tasks containing a major planning component (Shallice, 1982).

The results provide strong evidence for our second hypothesis, in which we predicted that a cueing procedure aimed at improving encoding should only improve the solving performance of frontal patients. Both left and bilateral frontal groups, performing significantly worse than normals in uncued conditions, were able to perform at normal levels after cueing. The reflection over individual sentences as well as over the schema of word problems improves the solving performance of both these frontal groups substantially. In contrast, the performance of patients with left posterior lesions even deteriorated slightly after cueing. This suggests that the problems left posterior patients encounter are of a different nature than the problems experienced by frontal patients.

Both groups' impairments are due to an inability to properly encode verbal information. However, our experiments demonstrate that the left and bilateral frontal patients' processing of text data is impulsive and haphazard, and can be improved by simply retelling problem sentences and clarifying the problem-schema. This cognitive cueing procedure does not improve the problem solving performance of patients with left posterior lesions. Moreover, these patients spend considerably more time in trying to solve arithmetic word problems, without any success in cued as well as in uncued conditions. These results provide indirect evidence for the finding that arithmetical word problem sentences and problem-schemata are far more difficult to understand and to represent for left posterior than for frontal patients (Luria & Tsvetkova, 1967). It also sustains the idea of Luria & Tsvetkova's (1967) that patients with left posterior lesions may get better help from material aids. One of these aids is a "dictionary", in which the most frequent logical grammatical expressions are written, along with their corresponding arithmetical operations.

Although patients with left and bilateral frontal lesions improved their performance after cueing, the absence of learning transfer from the first session to the second is striking. Such an absence of learning shown by patients with frontal lobe lesions argues in favour of a more general deficit in the spontaneous utilization of learning results to guide future responses. This deficit has been repeatedly noted in frontal lobe research (Luria, 1966; Lhermitte et al., 1972; Milner, 1982).

Patient groups with left and bilateral frontal lesions were the only ones in which solving time increased after cueing. The increase was only significant for patients with bilateral lesions. Patients with left frontal lesions were only fractionally slower after cueing. In other words, after cueing, patients with bilateral frontal lesions improve their performance at the expense of solving time, whereas with left frontal patients cueing only affects performance.

However, a word of caution is necessary. As opposed to the data on performance, solving times showed significantly greater variability in cued and uncued conditions, as measured by the standard deviation. Further inspection revealed two outliers in both the bilateral and the left frontal groups. A reassessment of the data of both groups, by excluding the aforementioned outliers on the basis of a 95 per cent confidence interval of the mean, suggests that in cued conditions the increase in solving time is also significant in the left frontal group (p=.05). This finding indicates the necessity to replicate the experiment with larger groups.

Finally, patients with left posterior lesions not only perform worse than patients with frontal lesions, but as was hypothesized, their solving times are also slower in every condition and do not change significantly from uncued to cued conditions. Only in cued conditions do patients with bilateral frontal lesions need as much time as patients with left posterior lesions to solve arithmetic word problems. These results support Luria & Tsvetkova's observation (1967) that, whatever the solving conditions, patients with left posterior lesions repeatedly try to understand what they are reading, spending long periods of time on this process without any result.

The findings of this study quite clearly indicate that a cognitive cueing procedure aimed at ameliorating sentence translation and problem-schema understanding improves the arithmetical word problem solving performance of left and bilateral frontal patients, but not of patients with left posterior lesions. However, two refinements of the experimental setup proposed in the present study may yield more precise indications about this improvement.

With regard to the cueing procedure, it might be argued that it acts simultaneously on two different encoding processes, i.e. sentence translation and problem-schema understanding, and that it does not allow to determine which of these processes contributes to the observed changes. Nevertheless, the aim of this study was to investigate the main stage of encoding, as distinct from the stage of memory-search in which a solution plan and adequate calculation procedures are devised. Moreover, our two former studies have shown that both sentence translation and problem-schema understanding are impaired in frontal as well as left posterior-injured subjects. At face value, however, and considering the reactions of surprise and disbelief following negative verbal feedback in problem-schemata cueing, these cues must have had only a slight effect on the improvements in problem solving performance. Further research with a more analytic procedure is needed to clear this therapeutically important issue.

The aim of the present cueing procedure was to reduce the patients' impulsivity and thereby enhance correct encoding. A question that remains to be answered is, of course, if impulsivity is a primary phenomenon causing an encoding deficit or if impulsivity is an indirect consequence of the absence of an encoding stage. In future research, the addition of a more neutral cueing condition, in which patients are asked to read word problems slowly and more carefully, may help to understand whether frontal patients are impulsive but still able to encode problem information or whether it is necessary to cue encoding directly.

CHAPTER VII

General Discussion and Concluding Remarks

The preceding chapters have put forward a study on the encoding stage of arithmetical word problem solving after frontal and left posterior brain lesions. In the first section of this final chapter, the new insights gained in the present investigation are discussed in the context of various theories on problem solving after frontal lobe damage. Subsequently, the theoretical and practical implications of the findings for further research are presented. Finally, the use of problem solving tasks characterized by a high predictive validity with regard to everyday problem solving is pleaded.

7.1. General Discussion

Through the use of a cognitive approach, the present study has sought to increase our understanding of impaired arithmetical word problem solving after frontal lobe damage. Within this cognitive approach of arithmetical word problem solving, the focus was on the stage of text-encoding. This stage consists of two distinct processes; sentence translation and integration of problem-information. In sentence translation, each word problem sentence is transformed into a semantic memory representation, whereas in the process of integration of problem-information the different sentences are put together in a coherent problem representation. The results of the first experiment performed in the present study reveal that, in comparison with healthy controls, both frontal and left posterior-injured patients are not able to represent word problem sentences correctly. In a sentence recognition task, both frontal and left posterior groups perform significantly worse than a group of healthy controls. In their turn, frontal patients perform significantly better than patients with left posterior lesions. Within the frontal group, patients with right frontal lesions perform better than patients with left and bilateral lesions, although the difference is only significant with regard to the right-bilateral groups comparison. In a sentence-picture matching task, healthy controls also perform better than frontal and left posterior patients. The difference between frontal groups and the left posterior group is not significant, suggesting that the left posteriors' group representation of arithmetical word problem sentences may be better preserved at a perceptual level than at a verbal level. These conclusions cast serious doubts on the interpretation of deficient arithmetical word problem solving as a mere defective recall of solving procedures (Barbizet, 1970). This interpretation presupposes that the information on which to operate is available to the frontal patient, but that he cannot remember how to proceed to a solution. The results of the present study, however, reveal that patients with frontal lesions operate upon erroneous and irrelevant problem information. Apparently, these patients proceed to a faulty solution mainly because they act on a defective representation of the problem givens and not only because their way of proceeding is impaired. In the same study, this viewpoint was confirmed when high error rates in the recognition of word problem sentences were found to be significant predictors for low arithmetical word problem solving ability.

A further study showed that the second text-encoding process, namely the integration of problem-information, is also impaired in frontal and left posterior-injured subjects. In this study, an arithmetical word problem sorting task revealed that healthy controls categorize arithmetic word problems according to principles that are basically different from those that guide the sorting behavior of frontal and left posterior patients. Healthy controls classify arithmetic word problems according to their "deep structure". In almost every case this "deep structure" coincides with the arithmetical operation governing the solution of the problem. Frontal and left posterior patients, on the other hand, use superficial problem features such as objects and arithmetical terms in order to sort these problems. This lack of recognition of a proper problem-schema can influence the solution of arithmetic word problems in two ways: it impedes an all ready-access to relevant problem solving procedures and it prevents the selection of relevant problem information. These results may explain Barbizet's description of frontal patients acting as if they had forgotten the correct way to proceed; these patients' surface-oriented categorization of arithmetic word problems does not yield any information that may be used to make inferences and derivations from the situation described by the problem statement and consequently does not permit to proceed to any kind of reasonable solution of the problem.

A third study revealed that a cognitive cueing procedure aimed at improving sentence translation and integration of problem-information improves word problem solving performance considerably with frontal, but not with left posterior patients. For both groups, this study confirms the vital importance of correct text-encoding for efficient arithmetical word problem solving. The results of the study also provide indirect evidence for the idea of Luria & Tsvetkova (1967), that the nature of the encoding deficit differs in both groups; retelling word problem sentences and clarifying the problem-schema apparently inhibits the impulsive encoding behavior of frontal patients, whereas it does not improve the defective understanding of problem-text in patients with left posterior lesions. With regard to theories that emphasize the role of the frontal lobe in planning and programming (Damasio, 1979; Shallice, 1982; Fuster, 1985), the results of the last study, suggest that an interpretation of defective arithmetical word problem solving after frontal lobe damage entirely in terms of a planning or programming deficit may be a bit restrictive. The conclusions of this last study, in combination with the findings of the other two studies, suggest that the defective encoding of problem-text has a decisive influence on arithmetical word problem solving performance. In information processing terms, this text-encoding stage precedes the stage of problem space search in which a solution program or plan is devised. The present findings suggest that in problem solving tasks with a substantial text-processing component, the encoding of the problem-text can be a major source of problems, even before planning impairments become apparent. This finding also throws a different light on two other specific frontal lobe symptoms, namely rule breaking and perseverative errors. The first of these symptoms concerns the compliance with task instructions. Subjects with frontal lobe lesions fail to carry out task instructions more frequently than patients with other cerebral lesions (Milner, 1965; Milner & Petrides, 1984). In accordance with our findings about impaired text-processing after frontal lobe damage, this "rule-breaking" behavior may be interpreted as an impulsive and haphazard encoding of verbal task instructions by frontal subjects. Another symptom frequently associated with frontal damage is the high incidence of perseveration errors (Cicerone et al., 1983; Milner & Petrides, 1984). These errors can derive from the inability to benefit from verbal feedback provided by the examiner. Patients with frontal lesions may encode these verbal cues so ineffectively that they do not guide future responses.

The results of the last study also revealed that frontal patients, particularly those with left and bilateral lesions, are not able to utilize a cueing procedure without external aid, suggesting an impairment in the spontaneous utilization of learning results. This is a finding that clearly fits into classic theories of frontal lobe dysfunctioning. The absence of spontaneous utilization of the cueing procedure and the presumably permanent necessity of external steering of encoding processes in frontal patients can be attributed to the absence of a Central Executive of memory (Baddeley, 1986) or to an impairment of a Supervisory System wich serves an executive function (Shallice, 1982).

The answer to the question whether the results of the present study add something new to the conclusions already formulated by Luria and Tsvetkova (1967) on the issue of arithmetical word problem solving after frontal lobe damage can be formulated at a methodological, a theoretical and a rehabilitational level. From a methodological viewpoint, the present study differs from Luria & Tsvetkova's in three aspects: (1) it investigates the issue of impaired arithmetical word problem solving with a controlled group design, (2) it includes a broader and better defined range of injuries and (3) IQ-levels and complexity of the word problems are controlled. At a theoretical level, this study clarifies what should be included in a debatable concept as "lack of orienting activity", used by Luria and coworkers to explain defective arithmetical word problem solving after frontal lobe damage. The present study demonstrates that such a concept should, if anything, include the idea of a defective arithmetical problem-text encoding. Finally, at a rehabilitation level, the findings of the present study supplement the ideas of Luria & Tsvetkova on the rehabilitation of impaired arithmetical word problem solving after frontal and left posterior brain damage. They suggest that the rehabilitation of patients with left and bilateral frontal lesions should include the learning of cognitive strategies aimed at properly encoding both the sentences and the problem-schema of arithmetic word problems.

7.2. Implications

The most interesting theoretical implication of the current study derives from the finding that defective arithmetical word problem solving after frontal lobe damage should not only be attributed to impaired planning or programming skills. This study suggests that before planning difficulties become evident, encoding problems can impair arithmetical word problem solving. A domain-specific analysis of arithmetical word problem solving led to this finding. In frontal lobe research investigators have frequently offered general theories in order to explain the functional role of the frontal lobes in thinking and problem solving. In other words, these investigators have emphasized impairments in one or more general frontal lobe functions (e.g. abstraction, planning, temporal integration) in order to explain defective processes of thinking or problem solving after frontal lobe damage. The results of the present study point to the limits of such a general approach. They indicate that these general approaches usually emphasize one or the other aspect of frontal lobe dysfunctioning, without fully accounting for the complex cognitive demands that problem solving tasks usually put on the solver. Moreover, these approaches often overlook the fact that different problem solving tasks may elicit a variety of different cognitive processes. Both these restrictions imply that in neuropsychological research more attention should be paid to the study of problem solving on the basis of task analyses and, hence, to a more cognitive approach of problem solving.

At a more specific level, the translation to an internal representation of word problem sentences appeared to be a major factor contributing to the impaired arithmetical word problem solving performance of both patients with frontal and left posterior brain lesions. A recognition and a sentence-picture matching task were used to demonstrate this effect. These tasks require sentence processing at a rather superficial level. Some of the hypothesized differences between the different groups and variables did, although present, not reach significance. Thus, both the hypothesized better verbal memory performance of patients with right frontal lesions in comparison with other frontal groups and the expected increase of recognition error rates from relation to compare sentences in the left posterior group were too slight to reach significance. In order to ascertain whether these differences really exist, research based on tasks requiring deeper processing than the simple recognition or the matching with pictures of arithmetical word problem sentences is needed.

With regard to the defective integration of problem-information into a useful problem-schema by patients with frontal as well as with left posterior lesions, it was concluded that this defective integration could affect arithmetical word problem solving performance. This conclusion was not reached directly, but was derived from research with expert and novice problem solvers (Larkin et al., 1980; Chi et al., 1981). In these investigations a relation between the recognition of adequate problem-schemata and the derivation of information relevant for word problem solving was established. It was also argued that the integration of problem-information based on "surface structures" could lead both groups to wrongly judge some superficial features of a word problem as being essential to its solution. The exact relation existing between deficient problem-schemata and word problem solving after frontal or left posterior brain damage, however, has still to be ascertained. Further research is required in order to answer questions like: which kind of procedural knowledge do brain damaged patients miss through inadequate problem-schemata, what are the inferences that brain damaged patients derive from their problem-schemata, how do problem-schemata influence word problem solving strategies in brain damaged patients, and so on.

One of the major goals of the current study was to investigate whether the arithmetical word problem solving performance of patients with frontal and left posterior brain lesions could be improved by using a cognitive cueing procedure. This cueing procedure was aimed at ameliorating sentence translation and problem-information integration. The results indicate that both encoding processes improve word problem solving in the left and bilateral frontal groups. An issue that remains to be addressed is that of the specific contribution of each of these cognitive processes to the ascertained performance gain in both frontal groups. An answer to this question requires that in future studies the influence of sentence translation and problem-information integration on arithmetical word problem solving performance should be investigated separately.

The implications for treatment derived from the present study can be quite plain. For patients with frontal brain damage, especially those with left and bilateral lesions, training has to focus on careful reading and encoding of problem-text, before any attempt is made to solve arithmetic word problems. Restating the problem givens and the problem goal in other words, representing problem sentences as pictures or diagrams, writing a short summary of the problem-text and comparing this summary with the original problem-text may be helpful exercises. Frontal patients must be made aware that they are "at risk" when they carelessly read and encode sentences containing much information, such as complex relation and compare sentences for example. The remedial treatment of word problem solving impairments in these patients should also include interventions aimed at understanding the problem-schema underlying arithmetic word problems. Useful exercises for this purpose may include the sorting of arithmetic word problems into categories, the recognition of relevant and irrelevant problem-information and the representation of word problems as equations or diagrams. Nonetheless, rehabilitation therapists should be aware that what frontal patients are taught about problem encoding, through sentence translation and information-integration exercises, is not always utilized spontaneously by these patients. Therefore, long-term mathematical education is required. Moreover, much remains to be done in the design and evaluation of compensatory cognitive strategies for the frontal patients' learning deficit.

On the other hand, patients with left posterior lesions appear to be in a different position. This study suggests that every attempt at improving their representation of problem sentences and problem-schemata by customary verbal means (e.g. repetition, clarification) is hardly effective. This suggests that the encoding deficit of these patients may have a more verbal character than the deficit of frontal patients. The use of external aids is the method of choice in order to help these patients in coping with their verbal encoding problem. Such an external aid can be offered in the form of a written instructional program. In this program various types of arithmetical word problem sentences may be registered along with their corresponding arithmetical operations or with their visual representation as, for example, diagrams. Without fully restoring the comprehension of arithmetical word problem sentences, the use of such a program could at least help the patient to compensate for his encoding impairment.

The design of rehabilitation programmes for frontal as well as for left posterior patients based on the principles mentioned above, the evaluation of their effectiveness and the examination of their impact on the patient's everyday arithmetical word problem solving behavior are as many topics for future research. The implementation of such retraining programmes is important because arithmetical word problem solving ability can have a considerable impact on everyday behavior, as will be explained in the next section.

7.3. Everyday Arithmetical Word Problem Solving

A prominent role for arithmetic word problems lies in their potential value as representations of real-world situations (Resnick & Ford, 1981). Such problems may offer the opportunity to utilize mathematical knowledge in a wide range of real-life situations, such as shopping, travelling and administering one's own financial affairs. The findings of neuropsychological research about the cognitive processes involved in solving and in learning to solve arithmetic word problems by brain damaged subjects may on the long term enhance the functional independence of these subjects. One could argue that context-impoverished school word problems, such as the problems used in the present study, could limit this value. Word problems with real-life story contexts are likely to serve this function better. However, for research purposes and as a starting point for understanding and revealing the nature of the cognitive processes involved in mathematical word problem solving, the use of standard problems seems to be preferable (Jaspers, 1991). Nevertheless, there is some evidence that the relationship between problem solving tests and everyday problem solving may not be a linear one. With age, for example, the performance in everyday problem solving may increase, whereas performance on traditional problem solving tests declines (Denney & Palmer, 1981; Cornelius & Caspi, 1987). Interestingly, neuropsychology has dealt with problem solving tests in some detail, but rarely with everyday problem solving. Moreover, there are hardly any neuropsychological studies in which the relationship between problem solving tests and everyday problem solving is examined. In order to bridge this gap, the development of tests which have a high predictive value with regard to everyday problem solving should receive strong attention. In the case of arithmetical word problem solving, research should be undertaken to investigate arithmetical word problem solving in everyday life and to compare everyday word problem solving with traditional indicators of arithmetical word problem solving ability. The assessment of empirical relations between both abilities will open up important new perspectives in both the diagnosis and the treatment of arithmetical word problem solving deficiencies after brain damage. After all, rehabilitation is not aimed at improving test-performance but at coping effectively with real life problems!

References

Alexander, M.P., Stuss, D.T., Benson, D.F. 1979. Capgras syndrome: A reduplicative phenomenon. In: Neurology, 29, 334-339.

Anderson, J.R. 1983. The Architecture of Cognition. Cambridge, Mass.: Harvard University Press.

Anderson, J.R. 1985. Cognitive Psychology and Its Implications. New York: W.H. Freeman and Company (2d. Ed.).

Ardila, A., Lopez, M.V. 1984. Transcortical motor aphasia: One or two aphasias ? In: Brain and Language, 22, 350-353.

Babinski, J. 1914. Contribution à l'étude des troubles mentaux dans l'hémiplégie organique cérébrale (Anosognosie). In: Revue Neurologique, 27, 845-848.

Baddeley, A.D. 1986. Working memory. London/New York: Oxford University Press.

Baddeley, A.D., Wilson, B. 1988. Frontal amnesia and the dysexecutive syndrome. In: Brain and Cognition, 7, 212-230.

Barbizet, J. 1970. Human Memory and its Pathology. San Francisco: W.H. Freeman & Company.

Bartus, R.T., Levere T.E. 1977. Frontal decortication in rhesus monkeys: A test of the interference hypothesis. In: Brain Research, 119, 233-248.

Benson, D.F., Gardner, H., Meadows, J.C. 1976. Reduplicative paramnesia. In: Neurology, 147-151.

Benson, D.F., Geschwind, N. 1975. Psychiatric conditions associated with focal lesions of the central nervous system. In: S. Arieti & M. Reiser (Eds.) American Handbook of Psychiatry, Vol. 4. New York: Basic Books.

Benton, A.L. 1968. Differential behavioral effects in frontal lobe disease. In: Neuropsychologia, 6, 53-60.

Benton, A.L. 1987. Mathematical disability and the Gerstmann syndrome. In: G. Deloche & X. Séron (Eds.) Mathematical Disabilities. A Cognitive Neuropsychological Perspective. Hillsdale, NJ: Lawrence Erlbaum Associates.

Berger, D.E. & Wilde, J.M. 1987. A task analysis of algebra word problems. In: D.E. Berger, K. Pezdek & W.P. Banks (Eds.) Applications of Cognitive Psychology: Problem Solving, Education, and Computing. Hillsdale, N.J.: Lawrence Erlbaum Associates.

Berlyne, D.E. 1965. Structure and Direction in Thinking. New York: Wiley.

Bianchi, L. 1895. The functions of the frontal lobes. In: Brain, 18, 457-522.

Bigler, E.D. 1988. Frontal lobe damage and neuropsychological assessment. In: Archives of Clinical Neuropsychology, 3, 279-297.

Blumer, D., Benson, D.F. 1975. Personality changes with frontal and temporal lobe lesions. In: D.F. Benson & D. Blumer (Eds.) Psychiatric Aspects of Neurologic Disease, Vol. 1. New York: Grune & Stratton.

Bobrow, D.G. 1968. Natural language input for a computer problem solving system. In: M. Minsky (Ed.) Semantic Information Processing. Cambridge, Mass.: MIT Press.

Bovenmeyer Lewis, A., Mayer, R.E. 1987. Students' miscomprehension of relational statements in arithmetic word problems. In: Journal of Educational Psychology, 79, 363-371.

Briars, D.J., Larkin, J.H. 1984. An integrated model of skill in solving elementary word problems. In: Cognition and Instruction, 1, 245-296.

Broca, P. 1861. Perte de la parole: Ramollissement chronique et destruction partielle du lobe antérieur gauche du cerveau. In: Bulletin de la Société d'Anthropologie, 2, 330-357.

Broca, P. 1865. Du siège de la faculté du langage articulé. In: Bulletin de la Société d'Anthropologie, 6, 377-393.

Buffery, A.W.H. 1967. Learning and memory in baboons with bilateral lesions of frontal or inferotemporal cortex. In: Nature, 214, 1054-1056.

Canavan, A.G., Janota I., Schurr, P.H. 1985. Luria's frontal lobe syndrome: psychological and anatomical considerations. In: Journal of Neurology, Neurosurgery and Psychiatry, 48, 1049-1053.

Canavan, A.G., Passingham, R.E., Marsden, C.D., Quin, N., Wyke, M., Polkey, C.E. 1989. Sequencing ability in parkinsonians, patients with frontal lobe lesions and patients who have undergone unilateral temporal lobectomies. In: Neuropsychologia, 27, 787-798.

Chi, M.T.H. 1978. Knowledge structures and memory development. In: R. Siegler (Ed.) Children's Thinking: What develops? Hillsdale, NJ: Lawrence Erlbaum Associates.

Chi, M.T.H., Feltovich, P.J., Glaser, R. 1981. Categorization and representation of physics problems by experts and novices. In: Cognitive Science, 5, 121-152.

Chi, M.T.H., Glaser, R., Rees, E. 1982. Expertise in problem solving. In: R. Sternberg (Ed.) Advances in the Psychology of Human Intelligence. Vol. 1. Hillsdale, NJ: Lawrence Erlbaum Associates.

Chi, M.T.H., Koeske, R.D. 1983. Network representation of a child's dynosaur knowledge. In: Developmental Psychology, 19, 29-39.

Chomsky, N. 1959. Review of B.F. Skinner's Verbal Behavior. In: Language, 35, 26-58.

Christensen, A.-L. 1975. Luria's Neuropsychological Investigation. New York: Spectrum Publications, Inc.

Cicerone, K.D., Lazar, R.M., Shapiro, W.R. 1983. Effects of frontal lobe lesions on hypothesis sampling during concept formation. In: Neuropsychologia, 21, 513-524.

Cooper, G., Sweller, J. 1987. Effects of schema acquisition and rule automation on mathematical problem-solving transfer. In: Journal of Educational Psychology, 79, 347-362.

Corkin, S. 1965. Tactually-guided maze learning in man: effects of unilateral cortical excisions and bilateral hippocampal lesions. In: Neuropsychologia, 3, 339-351.

Cornelius, S.W., Caspi, A. 1987. Everyday problem solving in adulthood and old age. In: Psychology and Aging, 2, 144-153.

Crowne, D.P. 1983. Frontal eye fields and attention. In: Psychological Bulletin, 93, 232-260.

Curtis, E.A., Jacobson, S., Marcus, E.M. 1972. An introduction to the neurosciences. Philadelphia: W.B. Saunders.

Damasio, A.R. 1979. The frontal lobes. In: K.M. Heilman & E. Valenstein (Eds.) Clinical Neuropsychology. New York: Oxford University Press.

Damasio, H. 1981. Cerebral localization of the aphasias. In: M.T. Sarno (Ed.) Acquired Aphasia. New York: Academic Press.

Damasio, A.R., Damasio, H., Chui, H.C. 1980. Neglect following damage to frontal lobe or basal ganglia. In: Neuropsychologia, 18, 123-132.

Damasio, A.R., Van Hoesen, G.W. 1980. Structure and function of the supplementary motor area. In: Neurology, 30, 359.

De Corte E., Verschaffel, L. 1985. Beginning first graders' initial representation of arithmetic word problems. In: Journal of Mathematical Behavior, 4, 3-21.

De Corte, E., Verschaffel, L. 1987. Using retelling data to study young children's word-problem-solving. In: J. Sloboda & D. Rogers (Eds.) Cognitive Processes in Mathematics. Oxford: Oxford University Press.

De Groot, A.D. 1946. Het denken van den schaker. Een experimenteel-psychologische studie. Amsterdam: Noord-Hollandsche Uitgevers Maatschappij.

Dellarosa, D. 1986. A computer simulation of children's arithmetic word problem solving. In: Cahier Research Methods, Instruments & Computers, 18, 147-154.

Deloche, G., Séron, X. (Eds.) 1987. Mathematical Disabilities. A Cognitive Neuropsychological Perspective. Hillsdale, NJ: Lawrence Erlbaum Associates.

Denney, N.W., Palmer, A.N. 1981. Adult age differences on traditional and practical problem-solving measures. In: Journal of Gerontology, 36, 323-328.

De Renzi, E., Faglioni, P., Lodesani, M., Vecchi, A. 1983. Performance of left brain-damaged patients on imitation of single movements and motor sequences. Frontal and parietal-injured patients compared. In: Cortex, 19, 333-343.

De Renzi, E., Vignolo, L.A. 1962. The Tokentest: a sensitive test to detect receptive disturbances in aphasics. In: Brain, 85, 665-678.

Drewe, E.A. 1975. An experimental investigation of Luria's theory on the effects of frontal lobe lesions in man. In: Neuropsychologia, 13, 421-429.

Ernst, G.W., Newell, A. 1969. GPS: A Case Study in Generality and Problem Solving. New York: Academic Press.

Faust, C. 1966. Different psychological consequences due to superior frontal and orbito-basal lesions. In: International Journal of Neurology, 5, 410-421.

Finan, J.L. 1942. Delayed response with pre-delay reinforcment in monkeys after the removal of the frontal lobes. In: American Journal of Psychology, 55, 202-214.

Fuster, J.M. 1980. The Prefrontal Cortex. Anatomy, Physiology, and Neuropsychology of the Frontal Lobe. New York: Raven Press.

Fuster, J.M. 1985. The prefrontal cortex, mediator of cross-temporal contingencies. In: Human Neurobiology, 4, 169-179.

Galperin, P.Y. 1969. Stages in the development of mental acts. In: M. Cole & I. Maltzman (Eds.) A Handbook of Contemporary Soviet Psychology. New York: Basic Books.

Glaser, R. 1984. Education and thinking: The role of knowledge. In: American Psychologist, 39, 93-104.

Goldstein, K. 1936. The significance of the frontal lobes for mental performance. In: Journal of Neurology and Psychopathology, 17, 27-40.

Goldstein, K. 1939. Clinical and theoretical aspects of lesions of the frontal lobes. In: Archives of Neurology and Psychiatry, 41, 865-867.

Goldstein, K. 1944. Mental changes due to frontal lobe damage. Journal of Psychology, 17, 187-208.

Goldstein, K. 1959. Functional disturbances in brain damage. In: S. Arieti (Ed.) American Handbook of Psychiatry. Vol. 1. New York: Basic Books.

Goldstein, K., Scheerer, M. 1941. Abstract and concrete behavior: An experimental study with special tests. In: Psychological Monographs, 53, 1-151.

Goodglass, H. Kaplan, E. 1972. The Assessment of Aphasia and Related Disorders. Philadelphia: Lea & Fabiger.

Grafman, J., Vance, S.C., Weingartner, H., Salazar, A.M., Amin, D. 1986. The effects of lateralized frontal lesions on mood regulation. In: Brain, 109, 1127-1148.

Greeno, J.G. 1974. Hobbits and orcs: Acquisition of a sequential concept. In: Cognitive Psychology, 6, 270-292.

Greeno, J.G. 1980a. Some examples of cognitive task analysis with instructional implications. In: R.E. Snow, P. Federico, W.E. Montague (Eds.) Aptitude, Learning, and Instruction. Vol. 2. Hillsdale, NJ: Lawrence Erlbaum Associates.

Greeno, J.G. 1980b. Trends in the theory of knowledge for problem solving. In: D.T. Tuma & F. Reif (Eds.) Problem Solving and Education: Issues in Teaching and Research. Hillsdale, NJ: Lawrence Erlbaum Associates.

Groen, G.J., Parkman, J.M. 1972. A chronometric analyis of simple addition. Psychological Review, 79, 329-343.
Grueninger, W.E., Pribram, K.H. 1969. Effects of spatial and nonspatial distractors on performance latency of monkeys with frontal lesions. In: Journal of Comparative and Physiological Psychology, 68, 203-209.
Guilford, J.P. 1967. The Nature of Human Intelligence. New York: McGraw-Hill.
Guitton, D., Buchtel, H.A., Douglas, R.M. 1982. Disturbances of voluntary saccadic eye-movement mechanisms following discrete unilateral frontal-lobe removals. In: G. Lennerstrand, D.S. Zee, E.L. Keller (Eds.) Functional Basis of Ocular Motility Disorders. Oxford: Pergamon Press.
Halstead, W.C. 1947. Brain and Intelligence: A Quantitative Study of the Frontal Lobes. Chicago: University of Chicago Press.
Hayes, J.R., Simon, H.A. 1974. Understanding written instructions. In: L.W. Gregg (Ed.) Kwowledge and Cognition. Hillsdale, NJ: Lawrence Erlbaum Associates.
Hayes, J.R., Waterman, D.A., Robinson, C.S. 1977. Identifying relevant aspects of a problem text. In: Cognitive Science, 1, 297-313.
Head, H. 1926. Aphasia and kindred disorders of speech. London: Cambridge University Press.
Hécaen, H. 1964. Mental symptoms associated with tumours of the frontal lobes. In: J.M Warren & K. Akert (Eds.) The Frontal Granular Cortex and Behavior, Ch. 16. New York: McGraw-Hill.
Hécaen, H. 1969. Aphasic, apraxic and agnosic syndromes in right and left hemisphere lesions. In: P.J. Vincken & G.W. Bruyn (Eds.) Handbook of Clinical Neurology, Vol. 4, Ch. 15. Amsterdam: North-Holland.
Hécaen, H., Albert, M.L. 1978. Human Neuropsychology. New York: John Wiley & Sons.
Hécaen, H.. Angelergues R. 1961. Étude anatomo-clinique de 280 cas de lésions rétro-rolandiques unilatérales des hémisphères cérébraux. In: Encéphale, 6, 533-562.
Hécaen, H., Angelergues, R., Houllier, S. 1961. Les variétés cliniques des acalculies au cours des lésions rétrorolandiques: Approche statistique du problème. In: Revue Neurologique, 105, 85-103.
Heilman, K.M., Valenstein, E. 1972. Frontal lobe neglect in man. In: Neurology, 22, 660-664
Heilman, K.M., Valenstein, E. 1979. Clinical Neuropsychology. New York: Oxford University Press.
Heller, J.I., Greeno, J.G. 1978. Semantic processing in arithmetic word problem solving. Paper presented at the Annual Meeting of the Midwestern Psychological Association, Chicago.
Henneman, E. 1980a. Organization of the motor systems: A preview. In: V.B. Mountcastle (Ed.) Medical Physiology, Vol.1, 14th. Ed. St. Louis: Mosby.
Henneman, E. 1980b. Motor function of the cerebral cortex. In V.B. Mountcastle (Ed.) Medical Physiology, Vol.1, 14th. Ed. St. Louis: Mosby.

Hinsley, D., Hayes, J.R., Simon, H. 1977. From words to equations: Meaning and representation in algebra word problems. In: P.A. Carpenter & M.A. Just (Eds.) Cognitive Processes in Comprehension. Hillsdale, NJ: Lawrence Erlbaum Associates.
Huber, W., Poeck, K., Willmes, K. 1984. The Aachen Aphasia Test. In: F. Clifford Rose (Ed.) Advances in Neurology. Progress in Aphasiology. New York: Raven Press.
Humphrey, G. 1951. Thinking. An Introduction to Its Experimental Psychology. London: Methuen & Co. Ltd.
Hussy, W. 1985. Komplexes Problemlösen – Eine Sackgasse? In: Zeitschrift für experimentelle und angewandte Psychologie, 22, 55-74.
Jacobsen, C.F. 1935. Functions of frontal association area in primates. In: Archives of Neurology and Psychiatry, 33, 558-569.
Jacobsen, C.F. 1936. Studies of cerebral function in primates. I. The functions of the frontal association areas in monkeys. In: Comparative Psychology Monographs, 13, 3-60.
Jaspers, M.W.M. 1991. Prototypes of Computer-Assisted Instruction for Arithmetic Word-Problem Solving. Doctoral Dissertation. Nijmegen: University of Nijmegen.
Johnson-Laird, P.N., Wason, P.C. 1977. Thinking. Readings in Cognitive Science. London: Cambridge University Press.
Kaczmarek, B.L.J. 1984. Neurolinguistic analysis of verbal utterances in patients with focal lesions of frontal lobes. In: Brain and Language, 21, 52-58.
Kertesz, A. 1979. Aphasia and Associated Disorders: Taxonomy, Localization and Recovery. New York: Grune & Stratton.
Kintsch, W. & Greeno, J.G. 1985. Understanding and solving word arithmetic problems. In: Psychological Review, 92, 109-129.
Köhler, W. 1921. Intelligenzprüfungen an Menschenaffen. Berlin: Springer.
Kolb, B., Milner, B. 1981. Performance of complex arm and facial movements after focal brain lesions. In: Neuropsychologia, 19, 491-503.
Kruskal, J.B., Wish, M. 1978. Multidimensional Scaling. Beverly Hills, California: Sage Publications.
Ladavas, E., Umiltà, C., Provinciali, L. 1979. Hemisphere-dependent cognitive performances in epileptic patients. In: Epilepsia, 20, 493-502.
Larkin, J. 1980. Teaching problem solving in physics: the psychological laboratory and the practical classroom. In: D.T. Tuma & F. Reif (Eds.) Problem Solving and Education: Issues in Teaching and Research. Hillsdale, NJ: Lawrence Erlbaum Associates.
Larkin, J. 1985. Understanding, problem representations, and skill in physics. In: S.F. Chipman, J.W. Segal & R. Glaser (Eds.) Thinking and Learning Skills: Research and Open Questions. Hillsdale, NJ: Lawrence Erlbaum Associates.
Larkin, J., Mc. Dermott, J., Simon, D.P., Simon, H.A. 1980. Expert and novice performance in solving physics problems. In: Science, 208, 1335-1342.
Leonard, G., Milner, B., Jones, L. 1988. Performance on unimanual and bimanual tapping tasks by patients with lesions of the frontal or temporal lobe. In: Neuropsychologia, 26, 79-91.

Leontiev, A. N. 1977. Probleme der Entwicklung des Psychischen. Athenäum Verlag, Kronberg/Ts.

Lezak, M.D. 1982. The problem of assessing executive functions. In: International Journal of Psychology, 17, 281-297.

Lhermitte, F., Derouesné, J., Signoret, J.L. 1972. Analyse neuropsychologique du syndrome frontal. In: Revue Neurologique, 127, 415-440.

Lishman, W.A. 1978. Organic Psychiatry. The Psychological Consequences of Cerebral Disorder. Oxford: Blackwell Scientific Publications.

Luria, A.R. 1966. Higher Cortical Function in Man. New York: Basic Books.

Luria, A.R. 1969. Frontal lobe syndromes. In P.J. Vincken & G.W. Bruyn (Eds.) Handbook of Clinical Neurology, Vol. 2, Ch. 23. Amsterdam: North-Holland.

Luria, A.R. 1973a. The Working Brain. An Introduction to Neuropsychology. New York: Basic Books.

Luria, A.R. 1973b. The frontal lobes and the regulation of behavior. In: K.H. Pribram & A.R. Luria (Eds.) Psychophysiology of the Frontal Lobes. London: Academic Press.

Luria, A.R., Karpov, B.A., Yarbuss, A.L. 1966. Disturbances of active visual perception with lesions of the frontal lobes. In: Cortex, 2, 202-212.

Luria, A.R., Tsvetkova, L.S. 1964. The programming of constructive activity in local brain injuries. In: Neuropsychologia, 2, 95-107.

Luria, A.R., Tsvetkova, L.S. 1967. Les troubles de la résolution de problèmes. Analyse neuropsychologique. Paris: Gauthier-Villars.

Luteijn, F., van der Ploeg, F.A.E. 1983. Handleiding Groninger Intelligentie Test (GIT). Lisse: Swets & Zeitlinger.

Maier, N.R.F., Burke, R.J. 1967. Response availability as a factor in the problem-solving performance of males and females. In: Journal of Personality and Social Psychology, 5, 304-310.

Malmo, R.B. 1942. Interference factors in delayed response in monkeys after removal of frontal lobes. In: Journal of Neurophysiology, 5, 295-308.

Maly, J., Quatember, R. 1980. Die Neuropsychologie Frontobasaler und Frontoconvexer Hirnläsionen. In: Zeitschrift für Klinische Psychologie und Psychotherapie, 28, 267-276.

Mandler, J.M., Mandler, G. 1964. Thinking: From Association to Gestalt. New York: John Wiley & Sons.

Mayer, R.E. 1982. Memory for algebra story problems. In: Journal of Educational Psychology, 74, 199-216.

Mayer, R.E. 1983. Thinking, Problem Solving, Cognition. New York: W.H. Freeman and Company.

Mayer, R.E. 1985. Mathematical ability. In: R. Sternberg (Ed.) Human abilities: An information processing approach. New York: W.H. Freeman and Company.

Mayer, R.E. 1986. Mathematics. In: R.F. Dillon & R.J. Sternberg (Eds.) Cognition and Instruction. New York: Academic Press.

Mayer, R.E. 1987. Learnable aspects of problem solving: Some examples. In: D.E. Berger, K. Pezdek & W.P. Banks (Eds.) Applications of Cognitive Psychology: Problem Solving, Education and Computing. Hillsdale, N.J.: Lawrence Erlbaum Associates.

Mayer, R.E., Larkin, J.H., Kadane, J. 1984. A cognitive analysis of mathematical problem solving ability. In: R. Sternberg (Ed.) Advances in the psychology of human intelligence. Hillsdale, N.J.: Lawrence Erlbaum Associates.

Mazzocchi, F., Vignolo, L.A. 1978. Computer assisted tomography in neuropsychological research: A simple procedure for lesion mapping. In: Cortex, 14, 136-144.

McFie, J. 1969. The diagnostic significance of disorders of higher nervous activity: Syndromes related to frontal, temporal, parietal and occipital lesions. In P.J. Vincken & G.W. Bruyn (Eds.) Handbook of Clinical Neurology, Vol. 4, Ch. 1. Amsterdam: North-Holland.

Mercer, B., Wapner, W., Gardner, H., Benson, D.F. 1977. A study of confabulation. In: Archives of Neurology, 34, 429-433.

Miller, G.A., Galanter, E., Pribram, K.H. 1960. Plans and the Structure of Behavior. New York: Holt, Rinehart & Winston.

Milner, B. 1964. Some effects of frontal lobectomy in man. In: J.M. Warren & K. Akert (Eds.) The Frontal Granular Cortex and Behavior, Ch. 15. New York: McGraw-Hill.

Milner, B. 1965. Visually-guided maze learning in man: effects of bilateral hippocampal, bilateral frontal, and unilateral cerebral lesions. In: Neuropsychologia, 3, 317-338.

Milner, B. 1982. Some cognitive effects of frontal-lobe lesions in man. In: D.E. Broadbent & L. Weiskrantz (Eds.) The Neuropsychology of Cognitive Function. London: The Royal Society.

Milner, B., Petrides, M. 1984. Behavioral effects of frontal-lobe lesions in man. In: Trends in Neurosciences, 7, 403-407.

Mohr, J.P., Pessin, M.S., Finkelstein, S., Funkenstein, H.H., Duncan, G.W., Davis, K.R. 1978. Broca aphasia: Pathologic and clinical. In: Neurology, 28, 311-324.

Morales, R.V., Shute, V.J., Pellegrino, J.W. 1985. Developmental differences in understanding and solving simple word problems. In: Cognition and Instruction, 2, 41-57.

Neisser, U. 1976. Cognition and Reality. San Francisco: W.H. Freeman and Company.

Nelson, H.E. 1976. A modified card sorting test sensitive to frontal lobe defects. In: Cortex, 12, 313-324.

Newell, A. 1980. Reasoning, problem solving, and decision processes: The problem space as a fundamental category. In: R. Nickerson (Ed.) Attention and Performance VIII, Hillsdale, N.J.: Lawrence Erlbaum Associates.

Newell, A., Simon, H.A. 1972. Human Problem Solving. Englewood Cliffs, N.J.: Prentice-Hall.

Novak, G.S. 1977. Representations of knowledge in a program for solving physics problems. Proceedings of the 5th International Joint Conference on Artificial Intelligence. Cambridge, Mass.: MIT Press.

Novoa, O.P., Ardila, A. 1987. Linguistic abilities in patients with prefrontal damage. In: Brain and Language, 30, 206-225.
Orbach, J., Fischer, G.J. 1959. Bilateral resections of frontal granular cortex: Factors influencing delayed response and discrimination performance in monkeys. In: Archives of Neurology, 1, 78-86.
Osgood, C.E. 1963. On understanding and creating sentences. In: American Psychologist, 18, 735-751.
Owen, E., Sweller, J. 1985. What do students learn while solving mathematics problems ? In: Journal of Educational Psychology, 77, 272-284.
Paige, J.M., Simon, H.A. 1966. Cognitive processes in solving algebra word problems. In: B. Kleinmuntz (Ed.) Problem Solving: Research, Method and Theory. New York: Wiley.
Pellegrino, J.W., Glaser, R. 1982. Analyzing aptitudes for learning: Inductive reasoning. In: R. Glaser (Ed.) Advances in Instructional Psychology. Vol. 2. Hillsdale, NJ: Lawrence Erlbaum Associates.
Polya, G. 1945. How to Solve It ? Princeton, NJ: Princeton University Press.
Pribram, K.H. 1961. A further experimental analysis of the behavioral deficit that follows injury to the primate frontal cortex. In: Experimental Neurology, 3, 432-466.
Pribram, K.H. 1967. The new neurology and the biology of emotion: a structural approach. In: American Psychologist, 22, 830-838.
Pribram, K.H. 1973. The primate frontal cortex - executive of the brain. In: K.H. Pribram & A.R. Luria (Eds.) Psychophysiology of the Frontal Lobes. London: Academic Press.
Resnick, L.B. 1976. Task analysis in instructional design: Some cases from mathematics. In D. Klahr (Ed.) Cognition and Instruction. Hillsdale, NJ: Lawrence Erlbaum Associates.
Resnick, L.B., Ford, W.W. 1981. The Psychology of Mathematics for Instruction. Hillsdale, NJ: Lawrence Erlbaum Associates.
Riley, M.S., Greeno, J.G. 1978. Importance of semantic structure in the difficulty of arithmetic word problems. Paper presented at the Annual Meeting of the Midwestern Psychological Association, Chicago.
Riley, M.S., Greeno, J.G., Heller, J.I. 1983. Development of children's problem-solving ability in arithmetic. In: H.P. Ginsburg (Ed.) The Development of Mathematical Thinking. New York: Academic Press.
Robinson, C.S., Hayes, J.R. 1978. Making inferences about relevance in understanding problems. In: R. Revlin & R.E. Mayer (Eds.) Human Reasoning. Washington: Winston/Wiley.
Ruff, R.L., Volpe, B.T. 1981. Environmental reduplication associated with right frontal and parietal lobe injury. In: Journal of Neurology, Neurosurgery and Psychiatry, 44, 382-386.
Rylander, G. 1939. Personality Changes after Operations on the Frontal Lobes. A Clinical Study of 32 Cases. Copenhagen: Munksgaard.
Salmaso, D., Denes, G. 1982. Role of the frontal lobes on an attentional task: a signal detection analysis. In: Perceptual and Motor Skills, 55, 127-130.

Schmidt, H.G., Norman, G.R., Boshuizen, H.P.A. 1990. A cognitive perspective on medical expertise: Theory and implications. In: Academic Medicine, 65, 611-621.
Selz, O. 1922. Zur Psychologie des produktiven Denkens und des Irrtums. Eine experimentelle Untersuchung. Bonn: Cohen.
Shallice, T. 1982. Specific impairments of planning. In D.E. Broadbent & L. Weiskrantz (Eds.) The Neuropsychology of Cognitive Function. London: The Royal Society.
Shapiro, B.E., Alexander, M.R., Gardner, H., Mercer, B. 1981. Mechanisms of confabulation. In: Neurology, 31, 1070-1076.
Simon, H.A. 1969. The Sciences of the Artificial. Cambridge, Mass.: The MIT Press.
Simon, H.A. 1978. Information processing theory of human problem solving. In: W.K. Estes (Ed.) Handbook of Learning and Cognitive Processes. Hillsdale, NJ: Lawrence Erlbaum Associates.
Simon, H.A. 1980. Problem solving and education. In: D.T. Tuma & T. Reif (Eds.) Problem Solving and Education: Issues in Teaching and Research. Hillsdale, NJ: Lawrence Erlbaum Associates.
Soloway, E., Lochhead, J., Clement, J. 1982. Does computer programming enhance problem-solving ability ? Some positive evidence on algebra word problems. In: R.J. Seidel, R.E. Anderson & B. Hunter (Eds.) Computer literacy. New York: Academic Press.
Spearman, C. 1927. The Abilities of Man. New York: McMillan.
Spiers, P.A. 1987. Acalculia revisited: Current issues. In: G. Deloche & X. Séron (Eds.) Mathematical Disabilities. A Cognitive Neuropsychological Perspective. Hillsdale, NJ: Lawrence Erlbaum Associates.
Stein, S., Volpe, B.T. 1983. Classical "parietal" neglect syndrome after subcortical right frontal lobe infarction. In: Neurology, 33, 797-799.
Sternberg, R.J. 1977. Component processes in analogical reasoning. In: Psychological Review, 84, 353-378.
Sternberg, R.J. 1983. Components of human intelligence. In: Cognition, 15, 1-48.
Sternberg, S. 1969. The discovery of processing stages: extensions of Donders' method. In: W.G. Koster (Ed.) Attention and Performance II. Amsterdam: North-Holland.
Stokx, L.C., Gaillard, W.K. 1986. Task and driving performance of patients with a severe concussion of the brain. In: Journal of Clinical and Experimental Psychology, 8, 421-436.
Stuss, D.T., Alexander, M.P., Lieberman, A., Levine H. 1978. An extraordinary form of confabulation. In: Neurology, 28, 1166-1172.
Stuss, D.T., Benson, D.F. 1984. Neuropsychological studies of the frontal lobes. In: Psychological Bulletin, 95, 3-28.
Stuss, D.T., Benson, D.F. 1986. The Frontal Lobes. New York: Raven Press
Stuss, D.T., Benson, D.F., Kaplan, E.F., Della Malva, C.L., Weir, W.S. 1984. The effects of prefrontal leucotomy on visuoperceptive and visuoconstructive tests. In: Bulletin of Clinical Neurosciences, 49, 43-51.

Stuss, D.T., Kaplan, E.F., Benson, D.F., Weir, W.S., Chiulli, S., Sarazin, F.F. 1982. Evidence for the involvement of orbitofrontal cortex in memory functions: An interference effect. In: Journal of Comparative and Physiological Psychology, 96, 913-925.

Teuber, H.-L. 1964. The riddle of frontal lobe function in man. In: J.M. Warren & K. Akert (Eds.) The Frontal Granular Cortex and Behavior. New York: McGraw-Hill.

Teuber, H.-L. 1966. The frontal lobes and their function: Further observations on rodents, carnivores, subhuman primates and man. In: International Journal of Neurology, 5, 282-300.

Thorndike, E.L. 1898. Animal intelligence: An experimental study of associative processes in animals. In: Psychological Review, 2, No. 8 (Monographical Supplement).

Thurstone, L.L. 1938. Primary Mental Abilities. Chicago: University of Chicago Press.

Tyler, H.R. 1969. Disorders of visual scanning with frontal lobe lesions. In: S. Locke (Ed.) Modern Neurology: Papers in Tribute to Derek Denny-Brown. Boston: Little, Brown & Company.

van Parreren, C.F. 1979. Het handelingsmodel in de leerpsychologie. Oration Francqui-Chair, University of Brussels.

van Parreren, C.F. 1983. Leren door handelen. Apeldoorn: Van Walraven.

Walsh, K.W. 1978. Neuropsychology. A Clinical Approach. Edinburgh: Churchill Livingstone.

Wang, P.L. 1987. Concept formation and frontal lobe function: The search for a clinical frontal lobe test. In E. Perecman (Ed.) The Frontal Lobes Revisited. New York: IRBN Press.

Watson, R.T., Heilman, K.M., Cauthen, J.C., King, F.A. 1973. Neglect after cingulectomy. In: Neurology, 23, 1003-1007.

Welch, K., Stuteville P. 1958. Experimental production of unilateral neglect in monkeys. In: Brain, 81, 341-347.

Wertheimer, M. 1945. Produktives Denken. Frankfurt am Main: Kramer.

Wertsch, J.V. 1981. Trends in soviet cognitive psychology. In: Storia e critica della psicologia, 2, 219-295.

Wickelgren, W.A. 1974. How to Solve Problems. Elements of a Theory of Problems and Problem Solving. San Francisco: W.H. Freeman and Company.

Wilkins, A.J., Shallice, T., McCarthy, R. 1987. Frontal lesions and sustained attention. In: Neuropsychologia, 25, 359-365.

Wundt, W. 1896. Grundriss der Psychologie. Leipzig: Engelmann.

Summary

This dissertation presents a cognitive neuropsychological approach to the issue of arithmetical word problem solving after frontal lobe damage. In six chapters, several aspects of this specific approach are highlighted. In the first chapter, a general introduction to the area of neuropsychology and arithmetical word problem solving is presented. The most influential theory related to this issue, namely A.R. Luria's, is briefly introduced. It is argued that a cognitive approach may offer a better perspective for the study of arithmetical word problem solving after frontal lobe damage.

The next two chapters discuss a number of issues pertaining to frontal lobe damage and arithmetical word problem solving. Chapter II gives a brief overview of the effects that frontal lobe lesions can have on human behavior, culminating into a description of the most influential theories dealing with the specific effects of frontal lobe damage on cognitive functioning. The difference between acalculia and defective arithmetical word problem solving is briefly discussed. Finally, several types of arithmetic word problems are outlined for which patients with frontal lesions show severe disturbances.

In chapter III, three different approaches to the issue of mathematical word problem solving after frontal brain damage are introduced and critically assessed. For several reasons, it is argued that the psychometric approach of mental abilities cannot provide a satisfactory explanation to the issue of word problem solving after frontal lobe damage. A second approach to mathematical word problem solving is based on the "theory of activity", developed within Russian psychology. After a general outline of this theory, its application by Luria and Tsvetkova to the issue of arithmetical word problem solving after frontal lobe damage is reviewed and discussed. It is concluded that some basic concepts of the "theory of activity", used to explain impaired word problem solving with frontal patients, are poorly defined and difficult to operationalize. Finally, a cognitive approach to the issue of arithmetical word problem solving is introduced. After that, a recent shift in emphasis within this approach is described; initially, cognitive analyses emphasized the role of general problem solving methods for skillful problem solving, whereas more recently the importance of domain-specific knowledge in problem solving has been highlighted.

Within this last framework, an information processing model of mathematical word problem solving was developed. The first main stage of this model, namely the encoding of arithmetical word problem-text, is the subject of the studies of this thesis. This text-encoding consists of two different cognitive processes: a translation process, in which each individual sentence of a word problem is transformed into a semantic memory representation, and an integration process, in which the different sentences of the word problem are put together in a coherent "problem-schema". Several arguments are put forward to justify the choice of the stage of text-encoding and thus of the processes of sentence translation and information-integration as a subject of study within arithmetical word problem solving. Finally, the method and the instruments used to investigate the processes of sentence translation and information-integration are presented.

Chapters IV, V and VI present the experimental studies which have been performed for this dissertation. In three separate studies, the questions raised in the previous theoretical chapters were examined.

In chapter IV sentence translation was investigated in right, left and bilateral frontally-injured patients; patients with left posterior lesions and healthy subjects served as controls. The translation of word problem sentences was examined with a recognition and a sentence-picture matching task. In addition, the relationship between sentence representation and arithmetical word problem solving was studied. The results suggest that the translation of sentences in the recognition, as well as in the sentence-picture matching task, is severely impaired in frontally lesioned patients, when compared with healthy controls. Within the frontal subgroups only one relevant difference was found: in the recognition task, patients with right frontal lesions performed significantly better than patients with bilateral frontal lesions, suggesting that verbal memory may be less affected by right frontal lesions. Patients with left posterior lesions made significantly more errors than frontally-injured patients and healthy controls in the recognition task, whereas no substantial difference between frontal and left posterior groups was found in the sentence picture-matching task. This suggests that patients with left posterior lesions may be slightly better at interpreting and understanding word problem sentences than at remembering them. Finally, a strong relationship was found between translation skills and arithmetical word problem solving capacity. This result emphasizes the major role of correct sentence translation in skillful arithmetical word problem solving performance.

In chapter V, a study is described in which a sorting task was used in order to examine the kinds of problem-schemata that healthy subjects, frontally and left posteriorly-injured patients impose upon arithmetical word problems of various types. Quantitative as well as qualitative analyses show that healthy subjects classify the arithmetical word problems according to principles that are essential for problem solving. These analyses also reveal that both the frontal and the left posterior brain-damaged groups base their sorting behavior on superficial text-characteristics, such as the objects referred to in the various word problems. The implications of these differences in sorting behavior for arithmetical word problem solving are discussed.

The purpose of the study reported in chapter VI is to examine whether a cueing procedure, aiming at the improvement of sentence translation and problem-schema understanding, can ameliorate arithmetical word problem solving in patients with frontal and left posterior brain lesions. The results indicate that this cueing procedure improves the solving performance of frontal patients but not of patients with left posterior lesions. This suggests that the nature of the encoding deficit may differ between these groups: the frontal patients' processing of problem-text seems to be impulsive and haphazard, whereas indirect evidence points to a language understanding deficit in patients with left posterior lesions. Although frontal patients improve their word problem performance significantly after verbal cueing, they are not able to utilize the cueing procedure spontaneously, suggesting an absence of a central executive or of a supervisory system which serves an executive function. The difference between both types of encoding deficit is discussed and some implications for rehabilitation are suggested.

In chapter VII the general findings of the present study are discussed in view of current theories about cognitive impairment after frontal lobe damage. It is concluded that, unlike some authors suggest, defective arithmetical word problem solving after frontal lobe damage cannot solely be attributed to a defective recall of solution procedures. It is also argued that in tasks with a substantial text-processing component, the correct encoding of the problem-text can be severely impaired for patients with frontal lobe lesions, even before impairments of planning become apparent. Several theories state that problems in planning are the main, if not the only, symptom characteristic to frontal lobe damage. The results of the present study show that these theories need to be refined. The same conclusion is drawn with respect to Luria and Tsvetkova's ideas. Next, several suggestions for further research are presented. The limitations of general theories of frontal lobe functioning are discussed and a more widespread use of task analysis in frontal lobe research is put forward. As a way of testing whether some slight effects found with the tasks of the present study are genuine, the use of tasks requiring a deeper level of processing is suggested. As some conclusions were based on studies with other than brain damaged subjects, a more direct approach of these topics is recommended.

The practical implications of the present findings, especially for rehabilitation, are twofold. Exercises for frontal patients should stress meticulous reading and encoding of the problem-text, whereas patients with left posterior lesions should use external aids enhancing the understanding of verbal information.

Finally, the use of problem solving tasks with a high predictive value with regard to everyday problem solving is advocated.

Samenvatting

Deze dissertatie beschrijft een cognitief-psychologische benadering van het oplossen van redactie-opgaven bij patiënten met frontale hersenletsels. In zes achtereenvolgende hoofdstukken worden diverse aspecten van deze specifieke benadering behandeld. In het eerste hoofdstuk wordt een algemene inleiding op het terrein van de neuropsychologische studie van redactie-opgaven gegeven. De meest invloedrijke theorie op dit gebied, die van A.R. Luria, wordt kort geïntroduceerd. Daarna worden enkele argumenten geformuleerd die pleiten voor een meer cognitief-psychologische benadering van het probleem van redactie-opgaven bij patiënten met frontale hersenlaesies.

In de twee volgende hoofdstukken komen enkele specifiekere aspecten van het frontaalsyndroom en het oplossen van redactie-opgaven aan de orde. Hoofdstuk II begint met een kort overzicht van de gevolgen die frontale hersenletsels kunnen hebben op het menselijk gedrag, gevolgd door een beschrijving van de meest invloedrijke theorieën die de specifieke gevolgen van frontale hersenletsels op het cognitief functioneren behandelen. Het verschil tussen acalculieën en stoornissen in het oplossen van redactie-opgaven wordt kort aangegeven. Tenslotte worden diverse typen redactie-opgaven beschreven die door patiënten met frontale hersenletsels niet kunnen worden opgelost. In hoofdstuk III worden drie mogelijke benaderingen van het probleem van redactie-opgaven bij patiënten met frontale hersenletsels geïntroduceerd en kritisch geëvalueerd. Op basis van diverse argumenten wordt gesteld dat een psychometrische benadering geen uitputtend antwoord kan bieden op het gestelde probleem. Een tweede benadering van het oplossen van redactiesommen na frontaal hersenletsel is gebaseerd op de zogenaamde "handelingstheorie", ontwikkeld in de Russische psychologie. Na een algemene beschrijving van deze theorie wordt de toepassing ervan door Luria en Tsvetkova op het onderwerp van redactie-opgaven na frontaal hersenletsel besproken. De conclusie is dat enkele basisbegrippen van de handelingstheorie moeilijk te definiëren en te operationaliseren zijn. Tenslotte wordt een cognitieve benadering van het probleem van het oplossen van redactie-opgaven na frontaal hersenletsel gepresenteerd. Er wordt ingegaan op een recente theoretische verschuiving binnen deze benadering: daar waar aanvankelijk bij cognitief-psychologisch onderzoek van probleemoplossen het belang van algemene oplossingsmethoden op de voorgrond stond, beklemtoont men sinds kort het belang van domeinspecifieke kennis.

Binnen dit domeinspecifieke kader is een informatieverwerkingsmodel van het oplossen van redactie-opgaven geformuleerd. Het eerste stadium van dit model betreft de verwerking van de opgavetekst en is het onderwerp van deze dissertatie. Deze verwerking van opgavetekst doet een beroep op twee verschillende cognitieve processen, te weten: een omzettingsproces, waarin iedere zin van de redactie-opgave wordt omgezet in een semantische geheugenrepresentatie, en een integratieproces, waarin de verschillende zinnen van de opgave worden samengevoegd tot een coherent probleemschema. Diverse argumenten worden gegeven waarom in dit proefschrift gekozen is voor de bestudering van het stadium van tekstverwerking, te weten van zinsrepresentatie en integratie van probleeminformatie. Tenslotte worden de methode en de instrumenten gepresenteerd die gebruikt worden om de representatie van zinnen en de integratie van probleeminformatie te onderzoeken.

In de hoofdstukken IV, V en VI worden de experimentele studies beschreven die voor dit proefschrift zijn verricht. In drie verschillende studies zijn de vragen, die in de voorgaande hoofdstukken werden gesteld, nader onderzocht.

In hoofdstuk IV is het hierboven genoemde zinsrepresentatieproces bij patiënten met rechtse, linkse en bifrontale hersenletsels onderzocht; patiënten met links-posterieure letsels en gezonde proefpersonen vormden de controlegroepen. De representatie van zinnen uit redactie-opgaven is onderzocht door middel van een herkenningstaak en een zin-figuur vergelijkingstaak. Bovendien is de relatie tussen de interne representatie van zinnen uit redactie-opgaven en het oplossen van deze opgaven onderzocht. De resultaten suggereren dat de representatie van zinnen uit redactie-opgaven in zowel de herkenningstaak als de zin-figuur vergelijkingstaak ernstig gestoord is bij patiënten met frontale hersenletsels. Binnen de diverse subgroepen met frontaal hersenletsel was er één significant verschil: in de herkenningstaak presteerden de patiënten met rechts-frontale letsels beter dan de patiënten met bilateraal-frontale letsels, hetgeen suggereert dat verbale geheugenprocessen minder ernstig gestoord zijn na rechts-frontale laesies. Patiënten met links-posterieure letsels presteerden significant slechter dan de frontale groep en de gezonde controles in de herkenningstaak, terwijl er geen significant verschil was tussen de links-posterieure en de frontale groepen in de zin-figuur vergelijkingstaak. Dit laatste suggereert dat patiënten met links-posterieure letsels zinnen uit redactie-opgaven enigszins beter kunnen interpreteren en begrijpen dan onthouden. Tenslotte wezen de resultaten op een sterke relatie tussen zinsrepresentatie en vaardigheid in het oplossen van redactiesommen. Dit laatste resultaat bevestigt de belangrijke rol van een adequate representatie van zinnen in het oplossingsproces van redactie-opgaven.

In hoofdstuk V wordt een studie beschreven waarin een sorteertaak werd gebruikt om na te gaan welke probleemschemata gezonde controles, patiënten met frontale letsels en patiënten met links-posterieure letsels gebruiken om verschillende redactie-opgaven te classificeren. Kwantitatieve en kwalitatieve analyses laten zien dat de gezonde proefpersonen de redactie-opgaven sorteren volgens principes die bepalend zijn voor de efficiëntie van het verdere oplossingsproces. Uit dezelfde analyses blijkt ook dat patiënten met frontale en patiënten met links-posterieure letsels hun sorteergedrag baseren op oppervlakkige, vaak fysieke kenmerken beschreven in het probleem, zoals bijvoorbeeld de voorwerpen die in het probleem worden genoemd. Aan het einde van de studie worden de implicaties van deze verschillen in sorteergedrag voor het oplossen van redactie-opgaven besproken.

Het doel van de in hoofdstuk VI beschreven studie is te onderzoeken of een verbale cueing-procedure om zinsrepresentatie en integratie van probleeminformatie te verbeteren ook tot een efficiëntere oplossing van redactie-opgaven leidt bij patiënten met frontale en links-posterieure hersenletsels. De resultaten tonen aan dat de cueing-procedure wel bij patiënten met frontale hersenletsels maar niet bij patiënten met links-posterieure letsels het oplossen van redactie-opgaven verbetert. Dit wijst erop dat de aard van de tekstverwerkingsstoornis in beide groepen verschilt: patiënten met frontale letsels coderen de opgavetekst op een impulsieve en willekeurige manier, terwijl er indirecte aanwijzingen zijn dat patiënten met posterieure letsels eerder met taalbegripsproblemen te maken hebben. Alhoewel de oplossingsvaardigheid van patiënten met frontale letsels significant verbetert met behulp van de cueing-procedure, blijken deze patiënten niet in staat de cueing-procedure spontaan te gebruiken, hetgeen wijst op een fundamenteel leerprobleem en op een stoornis in de organisatie van de taakuitvoering. De verschillen tussen de genoemde tekstverwerkingsstoornissen worden besproken en een aantal implicaties voor revalidatie worden aangegeven.

In hoofdstuk VII worden de bevindingen van de studies besproken in het kader van een aantal recente theorieën over cognitieve stoornissen na frontaal hersenletsel. Er wordt geconcludeerd dat stoornissen in het oplossen van redactie-opgaven na frontaal hersenletsel niet alleen kunnen worden toegeschreven aan het onvermogen om oplossingsprocedures op te roepen, zoals enkele onderzoekers stellen. Er wordt ook beargumenteerd dat in taken met veel verbale informatie bij patiënten met frontale hersenletsels de correcte representatie van de opgavetekst ernstig gestoord kan zijn, zelfs voordat er sprake is van enige planningsstoornis. Diverse theorieën stellen inderdaad dat stoornissen in de planningsvaardigheid het belangrijkste, zo niet het enige symptoom zijn dat karakteristiek is voor frontaal hersenletsel. De resultaten van de drie studies tonen aan dat deze theorieën gedeeltelijk herzien dienen te worden. Dezelfde conclusie geldt ook voor Luria's en Tsvetkova's ideeën. Vervolgens worden enkele suggesties voor verder onderzoek geformuleerd. De beperkingen van algemene theorieën over de rol van de frontaalkwab in het cognitief functioneren worden besproken en er wordt voorgesteld om de rol van taakanalyses in het onderzoek naar de gevolgen van frontaal hersenletsel uit te breiden. Ook wordt gesuggereerd om een aantal niet-significante verschillen opnieuw te onderzoeken met taken die een diepere tekstverwerking veronderstellen. Voor enkele conclusies, gebaseerd op studies met andere groepen dan hersenletselpatiënten, wordt een meer directe aanpak aangeraden.

De practische implicaties van de studies voor de revalidatie van stoornissen in het oplossen van redactiesommen zijn tweeledig. Voor patiënten met frontale hersenletsels worden oefeningen die een beroep doen op het nauwgezet lezen en verwerken van een opgavetekst aangeraden, terwijl bij patiënten met links-posterieure laesies het gebruik van materiële hulpmiddelen, die het begrip van verbale informatie ondersteunen, wordt geadviseerd.

Tenslotte wordt het gebruik van probleemtaken met een hoge voorspellende waarde ten aanzien van alledaagse probleemsituaties bepleit.